About Island Press

Since 1984, the nonprofit Island Press has been stimulating, shaping, and communicating the ideas that are essential for solving environmental problems worldwide. With more than 800 titles in print and some 40 new releases each year, we are the nation's leading publisher on environmental issues. We identify innovative thinkers and emerging trends in the environmental field. We work with world-renowned experts and authors to develop cross-disciplinary solutions to environmental challenges.

Island Press designs and implements coordinated book publication campaigns in order to communicate our critical messages in print, in person, and online using the latest technologies, programs, and the media. Our goal: to reach targeted audiences—scientists, policymakers, environmental advocates, the media, and concerned citizens—who can and will take action to protect the plants and animals that enrich our world, the ecosystems we need to survive, the water we drink, and the air we breathe.

Island Press gratefully acknowledges the support of its work by the Agua Fund, Inc., The Margaret A. Cargill Foundation, Betsy and Jesse Fink Foundation, The William and Flora Hewlett Foundation, The Kresge Foundation, The Forrest and Frances Lattner Foundation, The Andrew W. Mellon Foundation, The Curtis and Edith Munson Foundation, The Overbrook Foundation, The David and Lucile Packard Foundation, The Summit Foundation, Trust for Architectural Easements, The Winslow Foundation, and other generous donors.

The opinions expressed in this book are those of the author(s) and do not necessarily reflect the views of our donors.

State of the World 2012

MOVING TOWARD SUSTAINABLE PROSPERITY

Other Worldwatch Books

State of the World 1984 through *2011*
(an annual report on progress toward a sustainable society)

Vital Signs 1992 through *2003* and *2005* through *2011*
(a report on the trends that are shaping our future)

Saving the Planet
Lester R. Brown
Christopher Flavin
Sandra Postel

How Much Is Enough?
Alan Thein Durning

Last Oasis
Sandra Postel

Full House
Lester R. Brown
Hal Kane

Power Surge
Christopher Flavin
Nicholas Lenssen

Who Will Feed China?
Lester R. Brown

Tough Choices
Lester R. Brown

Fighting for Survival
Michael Renner

The Natural Wealth of Nations
David Malin Roodman

Life Out of Bounds
Chris Bright

Beyond Malthus
Lester R. Brown
Gary Gardner
Brian Halweil

Pillar of Sand
Sandra Postel

Vanishing Borders
Hilary French

Eat Here
Brian Halweil

Inspiring Progress
Gary T. Gardner

MOVING TOWARD SUSTAINABLE PROSPERITY

A Worldwatch Institute Report on Progress Toward a Sustainable Society

Erik Assadourian and Michael Renner, *Project Directors*

Jorge Abrahão	Colin Hughes	Mia MacDonald
Monica Baraldi	Paulo Itacarambi	Helio Mattar
Eric S. Belsky	Maria Ivanova	Monique Mikhail
Eugenie L. Birch	Ida Kubiszewski	Bo Normander
Robert Costanza	Henrique Lian	Michael Replogle
Robert Engelman	Diana Lind	Kaarin Taipale
Joseph Foti	Amy Lynch	Allen L. White

Linda Starke, *Editor*

ISLANDPRESS

Washington | Covelo | London

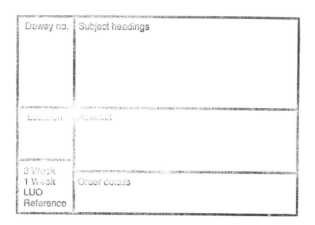

Dewey no.	Subject headings
Location	Abstract
3 Week 1 Week LUO Reference	Order details

ISBN 13: 978-1-61091-037-8
ISBN 10: 1-61091-037-0

The text of this book is composed in Galliard, with the display set in ScalaSans. Book design, cover design, and composition by Lyle Rosbotham.

✪ Printed on recycled, acid-free paper

Manufactured in the United States of America

Acknowledgments

Every year, Worldwatch Institute draws on the talents of hundreds of individuals and institutions from across the globe to assess the current state of the world. This year is no different. We want to express our deep gratitude to all those who helped *State of the World 2012* provide a fresh, new vision of sustainable prosperity as the world prepares to discuss this topic—and, indeed, humanity's survival—at the Rio 2012 summit this summer.

First and foremost, very special thanks to the Ford Foundation for supporting this year's report. We especially want to thank Don Chen, who proved to be a true champion of *State of the World* and connected us with leading experts whose ideas and prose fill many of the pages of the report. We also want to extend a warm thank you to Ford Foundation President Luis Ubiñas, who contributed an insightful Foreword for this year's report.

Early in the process of this year's project, a series of meetings by the Rio+20 Earth Summit Sustainable Cities Working Group helped a great deal in informing and shaping the report. A special thanks goes to all involved with those gatherings, especially Jacob Scherr of the Natural Resources Defense Council, who played a pivotal role in convening them.

Also, our gratitude to our Worldwatch colleagues Christopher Flavin and Gary Gardner for laying the foundations of this report and its underlying theme. Their initial vision carried this report even as the project evolved.

We offer a sincere thank you to our publishing partner Eduardo Athayde of Worldwatch Brazil. It was Eduardo's energy and commitment that connected us to many fine authors and partnership opportunities, which have been greatly appreciated.

Worldwatch Institute Europe, under the energetic leadership of Bo Normander, also played an important role in this year's report, contributing a chapter and several Boxes and adding to its online presence and the outreach of the findings. Worldwatch Europe's work was supported by the Velux Foundation of Denmark, for which all of us are grateful.

Every year our publishing partners play an important role in sharing the report far and wide, for which we are indebted. We in particular want to thank Gianfranco Bologna, who celebrates his twenty-fifth year as editor of the Italian edition. Gianfranco has played an important role in ensuring a strong publication, organizing presentations in Italy, and adding his vast knowledge of sustainability to *State of the World* for two and a half decades.

Thanks also to these many publishers: Universidade Livre da Mata Atlântica in Brazil; China Environment Science Press in China; Gaudeamus Helsinki University Press in Finland; Good Planet and Editions de La Martiniere in France; Germanwatch, Heinrich Böll Foundation, and OEKOM Verlag GmbH in Germany; Organization Earth in Greece; Earth Day Foundation in Hungary; Centre for Environment Education in India; World Wildlife Fund and Edizioni Ambiente in Italy; Worldwatch Japan; Africam Safari, Fundación Televisa, SEMARNAT, SSAOT (Secretaria de Sustentabilidad Ambiental del Estado de Puebla), and UDLAP in Mexico; Editura Tehnica in Romania; Center for Theoretical Analysis of Environmental Problems and International Independent University of Environmental and Political Sciences in Russia; Korea Green Foundation Doyosae in South Korea; Centre UNESCO de Catalunya for the Catalan version and CIP Ecosocial and Icaria Editorial for the Castilian version in Spain; Taiwan Watch Institute; Turkiye Erozyonla Mucadele, Agaclandima ve Dogal Varliklari Koruma Vakfi (TEMA), and Kultur Yayinlari Is-Turk Limited Sirketi in Turkey.

In 2012, Worldwatch is pleased to partner with Island Press as the publisher of the English-language version of *State of the World*. Since 1984, Island Press has been a trusted source of environmental information and solutions. We are delighted that *State of the World* will benefit from Island's network of digital and print distribution channels to ensure widespread availability of the ideas and proven strategies included in this edition. Thanks to their team—particularly David Miller, Brian Weese, Maureen Gately, Jaime Jennings, and Sharis Simonian.

Our readers are ably served by the customer service team at Direct Answer, Inc. We are grateful to Katie Rogers, Marta Augustyn, Colleen Curtis, Lolita Guzman, Cheryl Marshall, Ronnie Hergett, and Valerie Proctor for providing first-rate customer service and fulfilling our customers' orders in a timely fashion.

Thanks especially goes to our board, who guided us through challenging times this spring and helped us become a stronger organization. Many thanks also to the Institute's many individual and foundation funders, including Ray C. Anderson Family Foundation, Inc., The Bill & Melinda Gates Foundation, Barilla Center for Food & Nutrition, Climate Development and Knowledge Network (CDKN), Compton Foundation, Inc., Del Mar Global Trust, Ministry for Foreign Affairs of the Government of Finland, International Climate Initiative and the Transatlantic Climate Bridge of the German Federal Ministry for the Environment, Nature Protection and Nuclear Safety (BMU), Renewable Energy Policy Network for the 21st Century (REN21), The David B. Gold Foundation, Richard and Rhoda Goldman Fund and the Goldman Environmental Prize, Greenaccord International Secretariat, Energy and Environment Partnership with Central America (EEP), Hitz Foundation, Institute of International Education, Inc., Steven C. Leuthold Family Foundation, MAP Royalty, Inc. Sustainable Energy Education Fellowship Program, The Shared Earth Foundation, Shenandoah Foundation, Small Planet Fund of RSF Social Finance, V. Kann Rasmussen Foundation, United Nations Population Fund, Wallace Global Fund, Weeden Foundation, and the Winslow Foundation.

We also want to extend our gratitude to the Ethos Institute in Brazil, which generously supported this year's report and Brazilian outreach efforts in addition to contributing a chapter. The Ethos Institute works with companies to help them become more responsible and sustainable. Its involvement in this project was made possible by support to the Ethos Institute from several of the corporations—including Alcoa, CPFL Energia, Natura, Suzano, Vale, and Walmart Brazil—with which it developed a platform for an inclusive, green, and responsible economy. The platform, designed to govern corporate behavior in Brazil and beyond, will be an important step toward sustainable prosperity if it leads to a more responsible business sector.

Thanks to the entire Worldwatch Institute staff, who day in and day out make countless contributions to furthering the Institute's mission of a vision for a sustainable world. Thanks especially to Patricia Skopal Shyne, who is retiring after managing marketing and publication efforts for eight years. Patricia, who makes it look so easy to juggle infinite tasks, will be truly missed.

And every year the continuing patience and editorial sixth sense of *State of the World* esteemed elder and independent editor Linda Starke makes this whole process much easier. Designer Lyle Rosbotham also played a critical role in making this year's report engaging and readable. Artist Wesley Bedrosian, whose art is displayed on the report's cover, distilled the essence of the move toward sustainable prosperity perfectly.

We want to express our gratitude to the authors who contributed their expertise in chapters and in the many Boxes that expand the breadth and depth of the report. Thanks also to *State of the World* intern Matt Richmond who in the final months of the project helped us finish up the report.

We want to acknowledge the dozens of experts who helped strengthen the chapters this year and provided insights, data, and examples that help paint a fuller picture of the state of the world. While the names are too long to list here, we greatly appreciate their help.

The final thank you goes to all those who are working diligently to ensure that the upcoming Rio summit, and all opportunities in 2012 and beyond, will seek to ensure sustainable prosperity for all. Many have worked for decades—some even since the first global environment conference in Stockholm in 1972—to move humanity down a sustainable path. We offer our deepest thanks to these individuals. And thanks, too, to those brave individuals who pick up the reins as other advocates, reformers, and revolutionaries fall. To the next generation of activists, we give our thanks now—for if there is success in building a sustainable world, it will be due to your continuing energy and commitment.

Erik Assadourian and Michael Renner
Project Directors

Worldwatch Institute
1776 Massachusetts Ave, N.W.
Washington, DC 20036
www.worldwatch.org
www.sustainableprosperity.org

Contents

TABLES

FIGURES

Units of measure throughout this book are metric unless common usage dictates otherwise.

Foreword

Luis A. Ubiñas
President, Ford Foundation

Nearly a generation has passed since the Rio Earth Summit in 1992, and the world is now a vastly different place. An additional 1.5 billion people call our planet home. A majority of us now live in urban areas. A rapidly globalizing economy, massive waves of emigration and immigration, and revolutions in information technology mean we are all connected now more than ever.

But what, exactly, does all this mean for sustainable development? Rio+20 is a moment to answer that question—exploring how these rapid changes can be harnessed to advance sustainability and improve the lives of as many people as possible.

This edition of *State of the World* begins to do just that, and the Ford Foundation is proud to support it. This collection of fresh thinking, new tools, and provocative ideas shows us once again that a sustainable planet depends not only on the crucial decisions made at international conferences but also on innovation, energy, and commitment in our countless, ever-changing communities.

The following pages also make clear that the challenges before us are great if we are to foster a truly sustainable economy that advances human development today without sacrificing the human environment tomorrow. We've seen incredible progress—including greater formal recognition of the value of ecosystem

services, the rise of renewable forms of energy production, the development of market-based environmental management tools, and the adoption of sustainability practices in key sectors such as manufacturing and transport. But none of those actions have yet diminished the degradation of our shared environment. None of those actions have reduced the damage we are doing to our futures or the futures of our children and grandchildren.

Major questions remain about how a transition to a sustainable economy will take shape and whether such a shift will yield progress toward addressing a second scourge: the lives of poverty led by too many on this earth. For example, will green technologies offer opportunities for quality jobs and an improved standard of living in poor countries? Or will the economic benefits of such technologies be captured primarily by the wealthy and further widen the gap between rich and poor? Will recognition of the economic value of forests make it easier for rural and indigenous peoples to gain access to natural resources and pursue sustainable livelihoods? Or will it lead to new restrictions on land use by local communities? Will we take advantage of the rich cultural diversity of the world's traditional peoples? Or will their valuable heritage get washed away by globalization?

These are complex questions for which

there are no easy answers. But the ideas in this volume go a long way toward helping us map a path forward. They also reflect fundamental lessons that our partners across the world have again and again shown to be true—and that we believe are central to the sustainability agenda at Rio+20 and beyond.

First, it is abundantly clear that the active engagement of civil society is essential to the successful pursuit of the sustainability agenda. To fulfill the Rio+20 goal of poverty eradication through a green economy, civil society groups must be fully engaged. To that end, the Ford Foundation is supporting a wide range of organizations to voice their aspirations and concerns in the lead-up to the conference. We have also provided grants to international networks of advocacy groups, civil society institutions, and scholars working in key sectors such as housing, transport, and forest management. These stakeholders recognize that major economic transitions can present both opportunities and challenges for the working poor and for other marginalized people. We need their voices. Their active participation in the decisionmaking process will lend credibility to the next set of agreements to ensure that benefits are shared broadly and that negative consequences are carefully managed.

Second, we have seen time and again that empowering rural populations to act as stewards over natural resources holds tremendous value in the fight against climate change. The world's forests are not only home to hundreds of millions of people, they also are a key source of community livelihoods. For these individuals (many of whom are indigenous, tribal peoples), forests are a source of food, energy, medicine, housing, and income. Giving these communities the ability to own and manage the forests where they live provides perhaps the greatest incentive to protect and preserve these resources. Expanding community rights to forests—and other natural resources—is a working and successful model that many countries can and must follow.

Finally, it is clear that urban development and the tremendous growth of the world's cities must be central to any discussion of a sustainable future. The state of our cities is a pivotal issue that already touches the lives of half the world's population. And virtually all of the world's projected population growth over the next four decades—some 2.3 billion people—will take place in urban areas. Yet while some fret about rapid urbanization, we see tremendous possibility. The growth of cities can be an incredible opportunity for our collective efforts to expand economic opportunity, provide access to jobs and services that generate an income and build savings, gain social inclusion, and protect the environment. But to achieve these results, we need a fundamental mind-set shift: a new way of thinking about cities and urban development that embraces density, diversity, smart land use planning, and regularization. The way we collectively address urbanization will define the fate of billions of people and the sustainability of the planet.

The generations that follow ours—those of our children and our grandchildren—expect and need us to lead with wisdom and conviction today. They expect us to think not only of our time but of theirs, not only of ourselves but of them. As we mark the twentieth anniversary of the Rio Earth Summit with a new vision of a sustainable future, we have a chance to live up to our profound responsibility as stewards of the natural and man-made environments that sustain us. Let's make the most of this moment.

Preface

Robert Engelman
President, Worldwatch Institute

At times it seems that the only people who think United Nations environmental conferences have impact are those who distrust the United Nations and most things governmental. In the United States, at least, the news generated these days by *Agenda 21*—the agreement that emerged from the U.N. Conference on Environment and Development in Rio de Janeiro in 1992—is the conviction of some activists that the document represents a dangerous conspiracy to confiscate property and redistribute wealth. If you search YouTube for the phrase today you're more likely to encounter this incendiary thought than anything hopeful about the human future. As one who joined thousands of people from around the world in Rio to imagine an equitable and environmentally sustainable twenty-first century, I would find this amusing if it weren't so sad.

Go back even farther in time—twice as far, to 1972 and the first U.N. environmental conference in Stockholm—and the sense of wasted years is even more acute. Almost exactly 40 years before this book was published, environmental scientist Donella Meadows argued in *Newsweek* that the ethic of prosperity through endless economic and demographic growth would lead to a tragic reckoning on a finite planet. In 1972 there was no hint of the imminence of human-induced climate change or the end of cheap fossil fuels. Four decades

later, with the evidence of these all around us, the growth ethic still reigns.

And so in the weeks that followed the failed climate change conference in Copenhagen in 2009, when Worldwatch president Christopher Flavin suggested that we make the upcoming U.N. Conference on Sustainable Development (also known as Rio+20) the focus of *State of the World 2012*, I was dubious. Certainly the conference themes—jobs, energy, and food among them—were important and germane to the Institute's mission and work. But what do these meetings accomplish, I wondered, and how relevant are they even to readers interested in the environment?

One approach that helped convince me to forge ahead with Chris's idea, after I took over the leadership of Worldwatch in mid-2011, was to focus not so much on the conference itself as on the epic questions with which it will grapple. A dozen years into the twenty-first century we have little time left to bring the world's population—now 7 billion and counting—to a shared prosperity without bequeathing future humanity an overheated, resource-scarce, biologically impoverished planet. Yet even with the scientific evidence of our predicament now powerfully before us, governments have failed to develop policies that significantly limit environmental risk and spur equitable human development.

That distressing imbalance is reason enough to go—despite the cost in money, time, and (yes) carbon emissions—once more into the breach of environmental summitry. As I write, there has been little news media attention on the upcoming Rio gathering and no certainty that national leaders will attend. Even the activity among nongovernmental organizations is a fraction of the months-long whir of creativity I recall building up to the Rio Earth Summit of 1992. Yet as Jacob Scherr of the Natural Resources Defense Council points out, the conference will indeed take place. It will gather not just government delegates, development experts, and U.N. officials but thousands of citizen activists and other civil society representatives to ponder how a finite world can sustainably provide enough for all. And therein lies an opportunity—and much of the reason for this book's theme. With veteran *State of the World* project leaders Michael Renner and Erik Assadourian at the helm and new publisher Island Press behind the effort, we have aimed this year's book not so much at a city and a conference as at the fulcrum in history in which both feature.

At some point, greenhouse gas emissions will need to peak and begin falling. At some point, human fertility will need to fall below the level that spurs ongoing population growth. At some point, human development will need to reach thresholds at which all people can expect reasonable access to safe water, nutritious food, low-carbon energy, and decent health care, schools, and housing. After bold attempts in U.N. conferences to push governments toward strong action on the global environment and development in 1972 and 1992 (and at several points since), we can hope that the ideas for building sustainability have proliferated and ripened to the point where time and opportunity at last coalesce. We can hope that despite the many distractions and the pull of politics as usual, many in and out of government this year feel what Martin Luther King Jr. called, in a different but related context, "the fierce urgency of now" and can contemplate changing directions dramatically and fast.

The reports and ideas in the pages that follow are designed not as a blueprint for Rio's discussions but as proposals for that change, proposals to be considered and worked on before and after the conference ends. This book is the centerpiece of a wider Worldwatch project that will continue at least through 2012 to draw expanded attention and fresh ideas to the need for measurable action on green jobs, nutritious food, sustainable energy, safe water, healthy oceans, thriving cities, and fewer and less disruptive disasters—in short, to the need for shared prosperity worldwide that can be sustained for centuries to come. Keep an eye on our website, www.worldwatch.org, for more information, further articles, and word of upcoming conversations and related events, including launches of *State of the World 2012* in at least 20 languages by our many publishing partners around the world.

Most important, contribute your own energy and ideas to Rio+20 and the actions that follow after the delegates return home. Whatever presidents, parliaments, and parleys accomplish or do not accomplish, it is often social movements and citizen activists that spark the most momentous changes. This has been as true of the conservation and environment movements as it has been to the revolutions of civil and women's rights. Whatever the hour on the state of the world's environment and human development, there is hope and a long future ahead we will need to manage. We hope this book will take its place among a chorus of voices pointing the way.

Robert Engelman

State of the World: A Year in Review

Compiled by Matt Richmond

This timeline covers some significant announcements and reports from October 2010 through November 2011. It is a mix of progress, setbacks, and missed steps around the world that are affecting environmental quality and social welfare.

Timeline events were selected to increase awareness of the connections between people and the environmental systems on which they depend.

CONSUMPTION

WWF report finds that humans currently use 1.5 Earths, suggesting the world would need 50 percent more ecological capacity for current consumption patterns to be sustainable.

Biswarup Ganguly

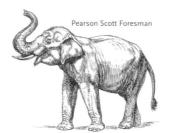

Pearson Scott Foresman

CLIMATE

Scientists discover microbes capable of eating hydrocarbons and natural gas, with potential to "fix" greenhouse gases deep in the ocean's crust.

ENDANGERED SPECIES

India declares the elephant a "National Heritage Animal" in order to increase protection of its 29,000 elephants.

MINING

In the Lima Declaration, Latin American indigenous tribes demand the end of large-scale mining in their territories.

OCTOBER   NOVEMBER

2010 STATE OF THE WORLD: A YEAR IN REVIEW

2 4 6 8 10 12 14 16 18 20 22 24 26 28 30   2 4 6 8 10 12 14 16 18 20 22 24 26 28

HAZARDOUS WASTES

General Motors agrees to set up a $773 million trust to clean up properties left behind after its bankruptcy, two thirds of which are contaminated with hazardous waste.

Andrew Jameson

Derelict GM Fisher body plant

Sander van der Molen

California oil refinery

GOVERNANCE

After 18 years of debate, 193 countries agree to a treaty defining how to cooperate in commercializing genetic resources.

CLIMATE

Overcoming a well-funded opposition campaign, Californians vote to keep the strongest greenhouse gas emissions standards in the United States.

NATURAL DISASTERS

A state of national catastrophe is declared in Colombia as intense rains affect at least 1.4 million people within the country's borders, killing more than 160.

ENERGY
New York becomes the first state to put a moratorium on hydraulic fracturing, a contentious form of natural gas drilling that can pose risks to drinking water supplies.

CLIMATE
Scientists find a "drastic" change in northern ocean currents that have a strong effect on weather and climate in the northern hemisphere.

ShacharLA

Dropping fire retardant

Canwest News Service

NATURAL DISASTERS
Israeli firefighters finally gain control of the worst wildfire in the nation's history, which burned over 10,000 acres and killed at least 42 people.

ENDANGERED SPECIES
The populations of four bumble-bee species in the US drop 96 percent, joining widespread losses in Europe and Asia of this important pollinator.

TOXICS
Study finds that 99–100 percent of expectant mothers have multiple highly toxic chemicals in their bodies, including mercury, PCBs, and flame retardants.

DECEMBER   JANUARY

2011

CLIMATE
Japan announces that it will not support the extension of the Kyoto Protocol past 2012, no matter how much pressure it faces.

NATIONAL SECURITY
US Senate approves a new strategic nuclear arms treaty with Russia, which would restart inspections of both nations' nuclear arsenals.

Martin Howard

CLIMATE
Australia cuts environmental programs in order to pay for massive flooding recovery, despite environmentalists' claims that climate change is behind the flooding.

TOXICS
The Archuar of Peru win an appeal to bring suit against Occidental Petroleum Corp. for 30 years of toxic wastewater dumping on their rainforested lands.

Mathew Brooks

Decommisioned Titan II

CLIMATE
NASA analysis finds 2010 ties 2005 as the warmest year on record.

Bernard Pollack

Goats in a Nairobi slum

HEALTH
International Livestock Research Institute details the dangers of livestock diseases crossing over to humans in the developing world, where new diseases emerge every four months.

BIODIVERSITY
Svalbard Global Seed Vault celebrates its third anniversary, with over 600,000 seed samples as a genetic-resource backup in the event of disaster.

NATURAL DISASTERS
A magnitude 9.0 earthquake and 10-meter tsunami waves devastate Japan, and the Fukushima Daiichi power plant suffers the worst nuclear disaster since Chernobyl.

GeoEye

TRANSPORTATION
The European Commission passes a long-term transport strategy that includes elimination of gas-powered cars in cities by 2050.

FEBRUARY   MARCH

2011   STATE OF THE WORLD: A YEAR IN REVIEW

2 4 6 8 10 12 14 16 18 20 22 24 26 28   2 4 6 8 10 12 14 16 18 20 22 24 26 28 30

ECONOMY
UN Environment Programme estimates that only 2 percent of world GDP is needed to transition the global economy toward sustainability.

Meena Kadri

POLLUTION
In an unprecedented ruling, Indian courts allow individuals to sue Coca-Cola for restitution based on environmental damage caused by their bottling plants.

TOXICS
Some 40,000 scientists and clinicians urge US federal agencies to go beyond current standards in assessing the safety of chemicals.

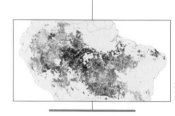

NASA

FORESTS
NASA mapping data shows over 1.3 million acres of the Amazon browned by record-breaking drought.

CULTURE
The Green Sports Alliance brings together teams in eight of the highest grossing US professional leagues to coordinate environmental initiatives.

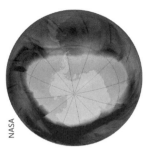

NASA

OZONE LAYER
Ozone loss over the Arctic reaches record levels, due to an especially cold winter in the stratosphere and the ozone-depleting substances still in the atmosphere.

ENERGY
Scientists at Los Alamos labs discover a cheap alternative to platinum in hydrogen fuel cells, a huge step toward reducing costs.

GOVERNANCE
Green Party wins its first seat in the Canadian House of Commons when its leader, Elizabeth May, is elected from British Columbia.

NATURAL RESOURCES
Wikileaks cables expose a "cold peace" among Arctic nations, all vying for the potential riches lying beneath the melting Arctic ice.

Shaun Merrit:

CLIMATE
International Energy Agency finds that emissions of energy-related CO_2 in 2010 were the highest in history.

APRIL   MAY

2 4 6 8 10 12 14 16 18 20 22 24 26 28 30   2 4 6 8 10 12 14 16 18 20 22 24 26 28 30

HEALTH
Nearly half of US meat and poultry is found to be contaminated with Staph bacteria, with 50 percent of it resistant to at least three known antibiotics.

NATURAL RESOURCES
Through the Law of Mother Earth, Bolivia grants to all of nature rights that are equal to humans.

CLIMATE
The Prince of Wales warns that ignoring climate change gives rise to the potential for a crash far more severe than the recent financial collapse.

ENERGY
German government announces that it will replace all 17 nuclear power plants with renewable energy sources by 2022.

USDA NRCS

Isofoton.es

Solar panels as highway noise barrier, Freising

Cylonka

TOXICS
Newfoundland joins Quebec,
Ontario, and New Brunswick in
banning all cosmetic pesticides
for residential lawns due to health
and environmental concerns.

Michelle Tribe

HEALTH
An EU law banning the sale
of baby bottles containing
Bisphenol A—
a potential endocrine
disruptor—goes into effect.

BIODIVERSITY
Lake Niassa, one of
the world's largest
and most biologically
diverse lakes, is
approved as a reserve
by the government
of Mozambique.

**NATURAL
DISASTERS**
Somalia and Eastern
Africa see their worst
drought in 60 years,
as tens of thousands
die of malnourishment
and 10 million more
need help to survive.

JUNE   JULY

2011 STATE OF THE WORLD: A YEAR IN REVIEW

2   4   6   8   10   12   14   16   18   20   22   24   26   28   30   2   4   6   8   10   12   14   16   18   20   22   24   26   28   30

TOXICS
Experts warn that
the sharp increase in
autism is likely due
in part to pregnant
women, fetuses,
and children being
exposed to a cocktail
of toxic chemicals.

Gray Watson

ENERGY
Google announces a $280-
million solar fund to help
homeowners buy solar
panels for residential use.

ECONOMY
Economists with the
Economics for Equity
and the Environment
Network find that each
ton of CO_2 emitted
causes up to $900 of
environmental harm.

John L. Alexandrowicz/EPA

HEALTH
Four months after the
Fukushima nuclear
meltdown, radiation
levels are up to 30
times above safe limits
for Japanese beef,
produce, and seafood.

ENERGY
Philips, the consumer electronics giant, wins the $10-million L-Prize from US Energy Department for a 9.7-watt LED light bulb, the equivalent of a 60-watt incandescent.

ENERGY
Exxon Mobil wins access to drill in Arctic waters off the Russian coast that are newly opened to exploration for oil.

WATER
After months of protests and violence, the government of Myanmar cancels construction of a dam on the Irrawaddy, the country's largest river.

LED torture testing

DOE

POLLUTION
Royal Dutch Shell manages to close a valve leaking oil into the North Sea, from which about 1,300 barrels had leaked in one week.

ENERGY
US National Center for Atmospheric Research finds that moving from coal to natural gas will actually increase the rate of global warming, calling into question this "bridge fuel."

Burma Democratic Concern

AUGUST   **SEPTEMBER**

2 4 6 8 10 12 14 16 18 20 22 24 26 28 30   2 4 6 8 10 12 14 16 18 20 22 24 26 28

NATURAL DISASTERS
US drought monitor reports that 73.5 percent of Texas is in "exceptional drought," the most severe category possible.

ENERGY
China begins allowing domestic solar power producers to sell their excess supply, hoping to build a domestic market for solar technologies.

© Bugwood.org
wooly adelgid

ENERGY
The US Army launches an initiative to generate 2.1 million megawatt hours of electricity through renewable energy sources.

US Army
National Guard solar carpark

Earl McGehee
Blanco River, Texas

BIODIVERSITY
Research finds that three invasive species—the emerald ash borer, gypsy moth, and hemlock woolly adelgid—cause $3.5 billion of damage annually in the US.

MARINE ECOSYSTEMS
Infectious salmon anemia, a virus deadly for salmon that evolved in Atlantic fish farms, is found for the first time in the Pacific Ocean.

GOVERNANCE
US Bureau of Labor Statistics data show that only 0.3 percent of layoffs in 2010 were due to government regulations, despite reports of "job-killing" regulatory regimes.

Utenriksdepartementet

Steve Vaughn

Corn ethanol plant, Iowa

FOOD
International Food Policy Research Institute claims that US corn ethanol subsidies are a major cause of the year's global food shortages.

POPULATION
Celebration in the Philippines as Danica May Camacho is born—one of a number of children chosen by the UN to be the world's symbolic 7 billionth inhabitant.

CLIMATE
The Durban climate negotiations open with South African President Jacob Zuma appealing to delegates to look beyond "national interests" for the good of humanity.

OCTOBER | NOVEMBER

2011 STATE OF THE WORLD: A YEAR IN REVIEW

2 4 6 8 10 12 14 16 18 20 22 24 26 28 30 | 2 4 6 8 10 12 14 16 18 20 22 24 26 28

BIODIVERSITY
An article in *Nature Climate Change* notes that global warming is shrinking not just the number but the actual size of many animal and plant species.

CLIMATE
Two hundred eighty-five of the world's largest investors urge governments to create a legally binding agreement on CO_2 emission reductions.

ENDANGERED SPECIES
Scientists warn that the Pacific yew tree, the main source of the chemotherapy drug Taxol, could soon be extinct due to overharvesting for medical purposes.

Eileen Beredo

Oyster farm, Washington State

MARINE ECOSYSTEMS
Report on massive die-offs of oyster larvae in the northwest US provides a glimpse into future effects of ocean acidification on marine life.

Walter Siegmund

See page 191 for sources.

MOVING TOWARD SUSTAINABLE PROSPERITY

CHAPTER 1

Making the Green Economy Work for Everybody

Michael Renner

In June 2012, Rio de Janeiro will host the United Nations Conference on Sustainable Development, more commonly referred to as Rio 2012 or Rio+20. The meeting marks the twentieth anniversary of the U.N. Conference on Environment and Development in 1992, also held in Rio. That landmark gathering adopted the Framework Convention on Climate Change and opened the Convention on Biological Diversity for signature. The conference was itself a milestone in the evolution of international environmental diplomacy, taking place two decades after the 1972 Stockholm Conference on the Human Environment.

On one level Rio 2012 marks a continuity of efforts to rally governments and civil society around the ever more urgent goal of reconciling human development with the limits of Earth's ecosystems. In 1992, the end of the cold war and rising environmental awareness seemed to open new horizons for global cooperation. The years since then have in many ways been a sobering experience, with sustainability aspirations often running headlong into discomforting political realities, orthodox economic thinking, and the staying power of materials-intensive lifestyles.

Among the obstacles to moving toward a more sustainable world order, writes Tom Bigg of the International Institute for Environment and Development (IIED), are "the interests of powerful constituencies that defend their turf and can manipulate the political system to stymie change; the hierarchy of policy and politics in almost every country which places environmental issues towards the bottom and economic growth and military security at the top; and the difficulty of achieving strong global regimes to effect change at a time when multilateralism is on the retreat."[1]

Environmental governance has largely taken a backseat to the pursuit of corporate-driven economic globalization—a process that has been marked by deregulation and privatization and thus a relative weakening of national political institutions. Comprehensive intergovernmental agreement on strategies for sustainability remains elusive. Despite multiplying numbers of solemn declarations, plans, and goals, no nation is even close to evolving toward a sustainable economy. The growth model that has emerged since the start of the Industrial Revolution, rooted in structures, behaviors, and activities that are patently unsustainable, is still seen as the ticket to ensuring the "good life"—

Michael Renner is a senior researcher at the Worldwatch Institute and co-director of *State of the World 2012*.

driven in no small measure by massive advertising. Western industrial countries hold fast to this model even in the face of rising consumer debt, while people elsewhere aspire to it.[2]

The Rio 2012 conference presents a much-needed opportunity to take stock of progress toward sustainability and development goals—and to create a new take on what prosperity means in the twenty-first century. Success will require not just official summitry but also imaginative initiatives to "lead from below" and qualitatively new relationships among governments, civil society, corporations, and the media.

A Complex Crisis

Humanity is confronting a severe and complex crisis. Mounting ecosystem stress and resource pressures are accompanied by growing socio-economic problems. The global economy is struggling to get out of a severe recession that was triggered by the implosion of highly speculative financial instruments but more broadly is the result of bursting economic bubbles and unsustainable consumer credit. The economic crisis is sharpening social inequities in the form of insecure employment and growing rich-poor gaps within and among countries.

All this has led to a growing crisis of legitimacy of economic and political systems, as massive bank bailouts stand in sharp contrast to austerity and curtailment of spending for the public benefit. The de facto appeasement of a run-amok financial system has blocked the emergence of a vision of how the real economy could be both rescued and made sustainable. Growing numbers of people sense that their interests are not represented in legislative and policymaking processes whose outcomes are increasingly influenced by money. Over the years, this has led to declining voter participation in elections and to political apathy.[3]

On the other hand, and more recently, disenchantment with the status quo has spawned

rapidly multiplying bottom-up protests now known as the "Occupy Movement." Before Occupy Wall Street was born, the "Indignados" (or Outraged) had camped out at the Puerta del Sol square in Madrid, and protesters took over public squares in Chile and Israel. The new movement derives some inspiration from the Arab Spring in the Middle East, suggesting a commonality of concerns across economic and political systems. The movement spread like wildfire. By mid-October 2011, Occupy protests had taken place in more than 900 cities around the world; by late December, there were activities in more than 2,700 locations.[4]

These protests have largely focused on social and economic concerns. But on the sidelines of the 17th Conference of the Parties (COP17) to the U.N. treaty on climate change that took place in Durban, South Africa, in December 2011, protesters made a connection to the fundamental issues of environmental sustainability. Organizers of Occupy COP17 argued that "the very same people responsible for the global financial crisis are poised to seize control of our atmosphere, land, forests, mountains and waterways." From Madrid to Manhattan to Durban, these actions are driven by deep frustration with the failure of governments and international conferences to address the fundamental problems that threaten human well-being and survival.[5]

In the two decades since the 1992 Earth Summit, pressures on the planet's natural resources and ecological systems have increased markedly as the material throughput of the economy keeps expanding. Not surprisingly, the bulk of human consumption is concentrated in cities. Urban areas account for half of the world's population but 75 percent of its energy consumption and carbon emissions.[6]

Ecological stress is evident in many ways—from species loss, water scarcity, carbon buildup, and nitrogen displacement to coral reef die-offs, fisheries depletion, deforestation, and wetlands losses. The planet's capacity to absorb waste and pollutants is increasingly taxed. Some 52 percent of commercial fish stocks are fully exploited, about 20 percent are overexploited, and 8 percent are depleted. Water is becoming scarce, and the supply is expected to satisfy only 60 percent of world demand 20 years from now. Although agricultural yields have increased, this has happened at the cost of declining soil quality, land degradation, and deforestation.[7]

A 2009 study of "planetary boundaries" showed that nine critical environmental thresholds had been crossed or were on track to be crossed, threatening to destabilize ecological functions on which economies, societies, and indeed all life on Earth critically depend. Humanity has been acting as if fresh resources were always waiting to be discovered, as if ecological systems were irrelevant to human existence, as if an Earth 2.0 were waiting in the wings in case we finally succeed in trashing this planet. There are isolated examples in human history of civilizations that outstripped their resource base, crashed, and vanished. But never before has this happened on a planetary scale; humanity is crossing into totally uncharted territory.[8]

While the impacts will be felt everywhere and especially in the poorest quarters, it is the actions of a minority that have gotten us to the edge of the precipice. According to the World Bank, people in the world's middle and upper classes more than doubled their levels of consumption between 1960 and 2004, compared with a 60 percent increase for those on the lower rungs of the income ladder. The global consumer class, about a billion people or so, mostly lives in western industrial countries, but the last two decades have witnessed the emergence of growing numbers of high consumers in countries like China, India, Brazil, South Africa, and Indonesia. Another 1–2 billion people globally aspire to the consumer life and may be able to acquire some of its trap-

pings. But the remainder of humanity—including the "bottom of the pyramid," the most destitute—have little hope of ever achieving such a life. The global economy is not designed for their benefit.[9]

Over the last decade, countries outside the Organisation for Economic Co-operation and Development (OECD) have increased their share of the world economy. From 40 percent of global gross domestic product (GDP) on a purchasing-power parity basis in 2000, their share has risen to 49 percent in 2010 and could grow to 57 percent by 2030. And economic expansion in countries like China, India, and Brazil has improved the economic lot of many people. According to OECD statistics, the number of poor people worldwide declined by 120 million in the 1990s and by nearly 300 million in the first half of the 2000s. And according to a World Bank analysis, the share of China's population earning less than $1.25 a day (in 2005 prices) dropped from 84 percent in 1981 to 16 percent in 2005. In Brazil the figures went from 17 percent in 1981 to 8 percent in 2005, and in India, from 60 to 42 percent.[10]

But it would be a mistake to regard the steady expansion of the global consumption intensive industrial economy as a surefire path toward overcoming poverty and social marginalization. The OECD notes: "The contribution of growth to poverty reduction varies tremendously from country to country, largely due to distributional differences within them. In many cases, growth has been accompanied by increased inequality." From 1993 to 2005 Brazil reduced poverty more than India did, even though its growth was much lower (1 percent versus 5 percent annually). This is because inequality has fallen in Brazil with the assistance of welfare programs like Bolsa Familia, but it has risen in China and India.[11]

Globalization has gone hand in hand with increased volatility and turbulence—and with great vulnerability for those unable to compete. The economic crisis that broke into the open in 2008 caused the ranks of the unemployed to swell from 177 million in 2007 to an estimated 205 million in 2010, with "little hope for this figure to revert to pre-crisis levels in the near term," the International Labour Organization (ILO) notes. Fears about "jobless growth" are borne out by an ILO analysis noting that the recovery of global GDP growth in 2010 was not paralleled by a comparable jobs recovery. And global emissions of carbon dioxide from fossil fuel burning rose by half a billion tons in 2010—the largest annual increase since the start of the Industrial Revolution. It is difficult to avoid the conclusion that the economy no longer works for either people or the planet.[12]

Even among those with a job, at least 1.5 billion persons worldwide—roughly half the workforce—are in highly vulnerable employment situations. The conditions they face—often referred to as "informality"—include inadequate or highly variable earnings, low-productivity work, temporary or insecure employment, and poor workplace conditions, especially in terms of occupational health and safety. Informal-sector workers typically earn about half as much as people in the formal sector.[13]

Rising numbers of people in industrial economies face precarious employment conditions as well. In the United States, wage stagnation and growing income inequality have been prominent phenomena since the late 1970s. Even though U.S. labor productivity expanded 80 percent between 1979 and 2009, average hourly compensation for workers rose just 8 percent, with most of the gains realized by the top earners. The number of Americans living below the official poverty line, about 46 million in 2010, is the highest in the 52 years since government statistics have been published on this topic. In Germany, long a high-wage country, the low-wage sector grew to more than 20 percent of

all employees as of 2008. In Japan, one third of the country's labor force is part-time and contract workers who lack job security. More than 10 million Japanese workers earn less than the official poverty line.[14]

There is a paradox. Wages are under pressure and employment is uncertain for many, yet consumerism remains alive and well. Materials-intensive lifestyles are financed not just by taking on additional jobs but also by going deeply into debt. The ILO explains that "in advanced economies, stagnant wages created fertile ground for debt-led spending growth—which is clearly unsustainable." In the United States in particular, high consumption was enabled by leveraging exaggerated housing values during the years of the real estate bubble.[15]

Worldwide, an extremely unequal distribution of wealth has emerged, with consequences for who has an effective voice in matters of economics and politics—and thus in how countries address the fundamental issues of sustainability and equity that confront humanity. A 2008 study by the UN University's World Institute for Development Economics Research (UNU-WIDER) offers data for the year 2000. (Data gaps and lags render a more up-to-date reckoning difficult.) The richest 1 percent of adults owned 40 percent of global assets. (See Figure 1–1.) For the top 5 percent, the share rises to 71 percent, and the top 10 percent controlled 85 percent of global wealth. By contrast, the bottom half of humanity together had barely even 1 percent of all wealth. The average member of the top 1 percent therefore was almost 2,000 times richer than the average person from the poorer half of humanity.[16]

It is unlikely that the last decade has brought a turn toward greater equality. Undoubtedly the regional distribution of wealth has undergone some shifts with the rise of countries like China, India, and Brazil. They now have a larger number of very wealthy individuals than in years past, and there is a rising middle

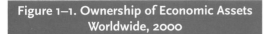

Figure 1–1. Ownership of Economic Assets Worldwide, 2000

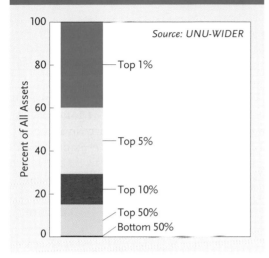

Source: UNU-WIDER

Top 1%

Top 5%

Top 10%

Top 50%
Bottom 50%

Green Growth and Degrowth

In times of economic crisis, environmental needs are quickly relegated to the status of a luxury. The conventional impulse is to "prime the pump" to get the economic engine moving again by whatever means necessary. Yet there is growing acceptance that the goals of environment and development are not necessarily in conflict. They can—and they need to—be reconciled. When governments reacted to the outbreak of the global economic crisis in late 2008, they did devote small portions of their economic stimulus efforts to a variety of "green" programs. Worldwide, an estimated 15 percent of stimulus funds went to support renewable energy and other low-carbon energy technologies, energy efficiency in buildings, low-carbon vehicles, and water and waste management efforts.[19]

class. But from a global perspective, these developments have not undone the observations from 2000 because, as the UNU-WIDER study documents, domestic wealth inequality is high in most countries.[17]

National data indeed suggest that inequality has been on the rise in many countries in recent years. In 2007, the richest 1 percent of Germans controlled 23 percent of the wealth in the country and the top 10 percent had 61 percent (up from 44 percent in 1998). The bottom 70 percent had just 9 percent. And in India, the top 1 percent had 16 percent of wealth in 2006; the top 10 percent had 53 percent. The bottom half of the population in India shared just 8 percent of the nation's wealth. In the United States, the share of wealth held by the top 5 percent increased from 59 percent in 1989 to 65 percent in 2009. The bottom 40 percent saw their net wealth fall from an already tiny 0.2 percent to a negative 0.8 percent. In fact, in 2009 almost a quarter of U.S. households had a zero or negative net worth, as consumer and mortgage debts cancelled or surpassed assets.[18]

In the face of crisis, new concepts such as a Global Green New Deal were developed. In the United Kingdom, the New Economics Foundation published a pioneering report on the topic, and the U.N. Environment Programme (UNEP) became a prominent advocate. UNEP also commissioned landmark reports on green jobs and the green economy.[20]

While the term "green economy" has gained currency, its meaning is still up for interpretation among governments, corporations, and civil society groups. UNEP defines a green economy quite broadly as one that results in "improved human well-being and social equity, while significantly reducing environmental risks and ecological scarcities. In its simplest expression, a green economy is low carbon, resource efficient, and socially inclusive." UNEP argues that "the greening of economies need not be a drag on growth. On the contrary, the greening of economies has the potential to be a new engine of growth, a net generator of decent jobs, and a vital strategy to eliminate persistent poverty."[21]

The extent to which a green economy and

economic growth are compatible is open to question, however. Developing technologies that are more resource-efficient and low-carbon is undoubtedly important and can help address some of the environmental problems humanity faces. But efficiency also makes consumption cheaper and may simply stimulate greater demand—a consequence that economists call the "rebound effect." Making a difference in the quest for sustainability will require an absolute decoupling of economic performance and materials use. (See Box 1–1.)[22]

The transition to a green economy is as much about social, political, and cultural change as it is about developing new technologies.

Mark Halle of the International Institute for Sustainable Development argues that a green economy "is not merely a redecoration of the traditional economy with green trimming, but a form of economic organization and priority-setting substantially different from the one that has dominated economic thinking in the richer countries for the past several decades."[23]

Because circumstances and needs vary so widely, industrial, emerging, and developing countries have different conceptions of what exactly a green economy entails—and how to get there. In fact, some observers in emerging and developing economies worry that green economy prescriptions could be used to justify

---

Box 1–1. The Role of Decoupling in a Green Economy

Decoupling human well-being from resource consumption is at the heart of the green economy. Typically, this is measured in terms of energy or materials use per dollar of gross domestic product. From 1981 to 2010, global energy intensity decreased by about 20 percent—or 0.8 percent each year. But this does not necessarily mean that growth in physical throughput and environmental impacts comes to an end. Indeed, during the same period world primary energy consumption expanded by 82 percent, from 6.6 billion tons of oil equivalent to 12 billion tons. Thus even an impressive rate of relative decoupling does not necessarily lead to an absolute decoupling.

This is also true for material throughputs. So the absence of even relative decoupling in the extraction of key metals like iron ore, bauxite, copper, and nickel is striking. Their consumption is rising faster than world GDP. If one day absolute decoupling of GDP from throughput becomes a reality globally, it will reinforce the logic of limiting throughput, providing evidence that environmentally costly resource use is no longer essential for generating wealth.

All this will need to change in the future. Fortunately there are signs that some countries may have already started down this decoupling path. Recent statistics show that in at least the United Kingdom absolute decoupling might have started a decade ago. In 2009, the country's total material requirement was 81 percent of its 2001 value.

If the idea of a green economy is to be taken seriously, the clear conclusion is that the world, starting with the most advanced countries, must engage in a discussion about a transition to "prosperity without growth." Making this possible requires a change in economic and social structures so that an economy without growth does not equal an unstable economy. One source of instability is clear: the wealthiest 20 percent of the world's population account for nearly 77 percent of total private consumption. Acceptance and implementation of prosperity without growth therefore requires a radical change—an immediate struggle against international and societal inequalities.

—José Eli da Veiga
University of São Paulo
Source: See endnote 22.

---

measures that block their developmental aspirations. A statement on behalf of the G77 nations cautions that a green economy "should not lead to conditionalities, parameters or standards which might generate unjustified or unilateral restrictions in the areas of trade, financing [official development assistance] or other forms of international assistance, leading to a 'green protectionism.'"A key challenge at the Rio 2012 conference is to address these worries, detailing the ways in which people in different parts of the world can derive benefit from a greener economy and committing to greater fairness in the distribution of resources and wealth.[24]

Figure 1–2, which brings together human development and ecological footprint information, shows that most countries are on either one or the other extreme of the spectrum: high development achieved on an unsustainable basis or low footprint at the cost of human deprivation. Only a smattering of countries come close to the "sustainable development quadrant."[25]

A green economy needs to be an appealing prospect. The aspiration is for "sustainable prosperity" for all—the result of a process of sustainable development that allows all human beings to live with their basic needs met, with their dignity acknowledged, and with abundant opportunity to pursue lives of satisfaction and happiness, all without risk of denying others in the present and the future the ability to do the same.

The world's consumer class needs to reduce its overconsumption—adjusting its focus from the accumulation of mostly short-lived, flimsy products that enter the waste stream at increasing speeds. Reducing its claim on resources would provide the ecological space needed to allow poor people to escape the deprivations of underconsumption. And considering that overconsumption has led to an obesity epidemic, social isolation, air pollution, traffic, and many other social ills, reducing consumption could have significant positive impacts on the well-being of the consumer class as well. Improving the lot of the world's poor would

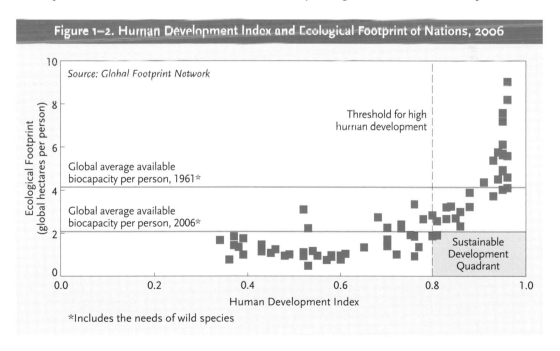

Figure 1–2. Human Development Index and Ecological Footprint of Nations, 2006

not have to come at the cost of a massive increase in carbon emissions. The *2011 Human Development Report* notes that providing everyone with at least basic modern energy services would increase emissions only 0.8 percent by 2030.[26]

The notion of a steady-state economy was examined by economist Herman Daly as early as 1973. Since then, many other studies and proposals have looked at how human well-being and happiness can be achieved without ever-increasing material throughput, be it in the form of making products more durable and repairable or work-time reductions and better sharing of work in line with greater productivity. With the passage of time, steady-state alone may no longer suffice. Some analysts argue that in order to live within the limits of Earth's capacity, the rich people of this planet need to undergo degrowth. (See Chapter 2.)[27]

Although the industrial countries bear major responsibility, Saleemul Huq of IIED argues that emerging economies may ultimately hold the key to a green economy. Undergoing massive economic growth, the emerging countries are starting to join the materialism of the old industrial countries. But they are not yet fully locked into a fossil-fuel-dependent economy and can leapfrog to technologies, structures, and lifestyles that are consistent with a low-materials "good life." Huq cautions that they will only do so if this is seen as a positive, pro-development opportunity rather than a burden urged on them. The Center on International Cooperation at New York University points out that emerging economies are not only "laboratories of the future" but also models that poorer developing countries might want to emulate.[28]

Developing countries have a major stake in the move toward a green economy. Already they confront the repercussions of the "brown economy" in the form of climatic upheaval. On average, natural resources and ecosystem ser-

vices provide about a quarter of the GDP in the poorest countries. In India, the poorest tenth of the population derives 57 percent of its GDP from ecosystem services through farming, animal husbandry, forestry, and fisheries. A continuation of current economic practices puts the natural assets on which the livelihoods, and lives, of hundreds of millions of poor people depend at increasing risk from climate change and other repercussions of ecological breakdown. More-sustainable and equitable provision of housing, transportation, energy, and sanitation could bring major benefits with regard to poverty reduction and healthier, safer lives.[29]

Establishing waste management and recycling operations that raise sanitary standards, for example, and providing clean drinking water and improved sanitation would substantially improve health and the quality of life, and it would generate much-needed employment. Decentralized provision of clean energy, including mini-grids and off-grid applications, can bring jobs and facilitate local business development.[30]

Growth of basic energy services, low-tech transportation networks, ecologically designed sanitation systems, and basic improved housing offer a double benefit: not only improving the daily lives of billions of people but also significantly reducing their ecological impacts. And these changes do not have to come at the expense of sufficient employment. To the contrary, they can contribute to more satisfying, meaningful livelihoods.

Green Jobs

One problem with the current economy is that it relies too much on limited and polluting resources such as fossil fuels and too little on an abundant resource—people. While greater labor productivity has undoubtedly been an engine of progress over time, its single-minded pursuit is turning into a curse.

From here on, progress requires a greater focus on energy, materials, and water productivity instead. Employment at adequate incomes is key to making an economy work for people, and therefore the transition to a green economy requires particular attention to good-quality jobs that contribute to preserving or restoring environmental quality.

For now, green jobs are still primarily found in a relatively small number of countries that lead in green R&D and investment, have adopted innovative pro-environmental public policies, and are able to build on strong scientific and manufacturing bases as well as on educated and skilled workforces. Countries like Japan, Germany, China, or Brazil already have the bulk of employment in renewable energy, energy and materials efficiency, and related fields. But growing numbers of countries are claiming a share of the green economy. And employment in installing, operating, or maintaining equipment like solar panels, wind turbines, insulation materials, rail vehicles, or efficient industrial equipment will be more widely spread than jobs in green manufacturing.

A sustainable economy requires social solidarity and equity between and within countries and cannot be built on "green for a few" policies—with benefits for only some countries, some companies, or some workers. Instead, there is a need for a "green for all" strategy, with new approaches in energy provision, transportation, housing, and waste management that combine technical and structural change with social empowerment.

Energy. Energy use pervades virtually every human activity on Earth, and the heavy reliance on fossil fuels is a major culprit behind urban air pollution and climate change. In 2010, oil, gas, and coal accounted for 87 percent of commercial primary energy use. Renewables (including hydropower) contributed 8 percent, and nuclear energy, 5 percent. But many people in developing countries contend with energy poverty—suffering from inadequate access to energy in general and relying on traditional, polluting biomass (firewood, charcoal, manure, and crop residues).[31]

A green and equitable energy transition will require richer individuals to both switch from fossil fuels and reduce their energy demand via greater efficiency and conservation efforts, whereas the poor will require more, and cleaner, energy. Both dimensions of this transition offer employment opportunities. On the whole, the energy sector is a relatively small employer, notwithstanding its catalytic effect on the entire economy. But renewables tend to be more jobs-intensive than the already highly automated, mature fossil fuel industry, and the pursuit of energy efficiency similarly offers greater job opportunities than increasing energy supply does.[32]

Renewable energy is expanding fast. From just $7 billion in 1995, global investments surged to $243 billion in 2010, principally in wind energy ($96 billion) and solar power ($89 billion). In terms of total renewable power installed (excluding hydropower), the leaders are the United States, China, Germany, Spain, and India. (If hydropower is included, Canada and Brazil join the ranks.) Figure 1–3 provides details of installed capacity in wind power, solar photovoltaics (PV), and solar heating, as well as biofuels production.[33]

In 2010, wind energy represented by far the largest chunk of renewable power generating capacity in the world, followed by biomass power and solar PV. The latter is picking up speed, with global capacity growing at an average annual rate of 49 percent between 2005 and 2010, compared with 27 percent each for wind power and concentrating solar power and 16 percent for solar hot water. Bioethanol production expanded 23 percent annually and biodiesel, 38 percent.[34]

More than 100 countries are now developing wind power capacities. The leading wind turbine manufacturers are based in China,

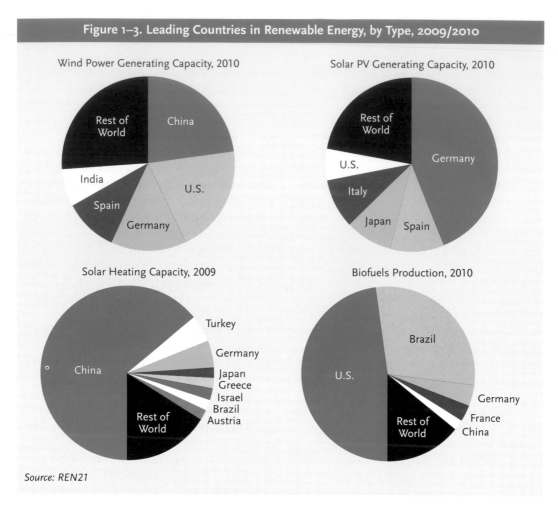

Figure 1–3. Leading Countries in Renewable Energy, by Type, 2009/2010

Wind Power Generating Capacity, 2010

Solar PV Generating Capacity, 2010

Solar Heating Capacity, 2009

Biofuels Production, 2010

Source: REN21

Denmark, Germany, the United States, Spain, and India. By capacity installed, the global leaders are China, the United States, Germany, Spain, and India. As Spain's Navarra region has demonstrated, the development of wind power can bring substantial local benefits. Navarra, which now derives two thirds of its electricity from renewables, managed to cut unemployment from a peak of 12.8 percent in 1993 to 4.8 percent in 2007—the result of an active industrial policy intended to build wind power capacity and a concerted worker training effort focused on this industry.[35]

Companies based in China, Taiwan, the United States, Germany, and Japan are the global leaders in manufacturing solar PV panels. But even in countries with no domestic solar manufacturing industry there are important job opportunities in sales, assembly and installations, and maintenance. Small solar PV systems already provide power to a few million households in developing countries, and solar cookers and solar portable lights offer a range of benefits. In Bangladesh, micro-credit schemes helped spread solar systems from 320,000 homes in 2009 to 1.1 million by August 2011.[36]

Biofuels production is expanding, although

controversy continues to rage over food-versus-fuel issues and whether such fuels offer a net carbon benefit compared with fossil fuels. Ethanol and biodiesel together provided about 2.7 percent of worldwide road fuels in 2010. Brazil has by far the largest bioethanol industry. About half a million people work in sugarcane cultivation for biofuels use, and another 190,000 in processing the sugarcane into ethanol. Biogas is also growing in significance, with more than 44 million households worldwide relying for their lighting and cooking needs on community- or household-scale biogas digesters. China leads the world, but gasifiers for heat generation are also increasingly used in India and other countries.[37]

Although data on employment are not systematically collected and gaps persist, the number of renewable energy jobs worldwide is undoubtedly rising fast. A rough estimate suggests at least 4.3 million direct and indirect (that is, supply chain) jobs, up from an estimate of 2.3 million in 2008. These estimates are incomplete and do not fully account for the jobs or livelihoods in connection with many rural energy projects.[38]

Renewable energy employment is still smaller than fossil fuel employment. The extraction of oil, gas, and coal employs more than 10 million people, and the use of these energy sources in thermal and electricity plants adds several million more jobs. But given that renewable energy still accounts for a small share of total energy use, the number of people already working in this field is encouraging.[39]

Transportation. The transportation sector, especially the close to 1 billion motor vehicles on the world's roads, accounts for more than half of global liquid fossil fuel consumption. Accounting for about a quarter of energy-related carbon dioxide emissions, with emissions rising faster than those of any other economic sector, transportation is an important contributor to climate change. Its other impacts include urban air pollution, accidents,

congestion, noise pollution, and obesity. Social dimensions deserve equal attention: where dependence on private automobiles is heavy and public transport options are sparse or non-existent, it can be expensive, and perhaps impossible, for people to secure access to jobs and livelihoods without cars. (See also Chapter 4.)[40]

Efforts to reduce transportation's footprint have principally focused on technology—measures to boost vehicle fuel efficiency, switch to alternative fuels, and develop hybrid and electric vehicles. Although automobile fuel efficiency has been improving in recent years, truly efficient models still do not come close to even one tenth of total sales, and hybrid and electric vehicles presently account for less than 3 percent.[41]

A number of countries are putting their faith in the development of biofuels. Brazil is now producing almost exclusively "flex-fuel" vehicles that can run on any blend of gasoline and ethanol, and it is hoping to convert its entire fleet over the next 20 years or so. More than 80 countries, many of them poor, have decided to pursue a different alternative: vehicles running on natural gas (mostly compressed natural gas, or CNG), which burns more cleanly than gasoline. Pakistan, Iran, Argentina, Brazil, and India accounted for three quarters of the global CNG fleet of close to 13 million in 2010.[42]

But such measures alone are inadequate in the face of growing numbers of vehicles and longer distances driven. Rich countries in particular need to reduce their heavy car dependence. Other countries, too, are already emulating or aspiring to build an automobile-centric system, often at the cost of badly polluted and congested cities. Especially in poor societies, public spending in support of car-centric transportation systems accentuates social disparities. Expenditures on roads crowd out other needed public infrastructure and marginalize those who cannot afford a car.

In both wealthy and poor countries, a reliable and affordable public transit system plays a critical role in achieving a greater degree of social equity. Poorly planned or designed transport systems and unnecessary sprawl can make access to jobs physically difficult and costly, especially for low-income households in both rich and poor countries, which have to allocate a disproportionate share of their meager incomes to cover transport expenses.

A more forward-looking policy seeks to achieve a better balance of transportation modes and thus to boost the share of public transit in cities and rail in intercity travel. By avoiding sprawl and limiting the distances that must be traveled by people and freight, options like public transit, biking, and walking become more feasible.

Such changes have implications for the transportation sector workforce. While no truly comprehensive studies have been undertaken on the employment implications of a far-reaching modal shift, some rough figures indicate the current situation. Direct employment in manufacturing motor vehicles runs to more than 8 million people worldwide, with multiples of this figure in the supply chain. By comparison, relatively few people are today employed in manufacturing rail vehicles—about half a million directly. Larger numbers of people work in operating public transportation systems: more than 7.6 million in urban mass transit and 7.1 million in freight and passenger railways.[43]

There are some encouraging changes under way, all of which are translating into increased employment in operating public transportation systems. Ridership in urban transit and intercity rail is rising worldwide, as are investments in these transportation systems. Interest in high-speed rail is growing around the world. Japan, France, Spain, and China are at the forefront, but the number of countries running such trains is expected to grow from 14 in mid-2011 to 24 over the next few years. Bus Rapid Transit systems on a broad scale were pioneered in Curitiba, Brazil, in 1974. Particularly since the 1990s, this concept is spreading to a growing numbers of cities. By 2005, an estimated 70 Bus Rapid Transit systems were in operation worldwide.[44]

Buildings. Approximately one third of global energy end-use takes place within buildings, and nearly 60 percent of the world's electricity is consumed by residential and commercial buildings. Under business-as-usual assumptions, building energy demand is projected to increase by 60 percent by 2050. Yet this sector also offers enormous potential for significant energy savings and carbon emissions reductions through more-appropriate building materials and greater insulation in windows and roofing, as well as reliance on more-efficient heating and cooling systems, lighting, appliances, and equipment in buildings.[45]

The construction industry also carries great importance as an employer. In most countries, it accounts for anywhere from 5 to 10 percent of all jobs, though often with strong seasonal variations. Worldwide, at least 111 million people find work in this sector. But given that the industry is highly fragmented and that many workers are in informal employment arrangements that evade capture in official statistics, the real figure is likely to be much higher.[46]

The renovation and retrofitting of existing buildings tends to be of greater importance in industrial countries with a large existing building stock and low population growth rates. In developing countries, in contrast, greening new construction is very important, especially in China and India, where the economies are expanding fast and rural residents stream into cities in search of work. In the developing world, informal and often substandard housing is widespread; improving health and safety standards there is as much of an urgent task as greening the building stock is.

The share of the urban population living in

slums in the developing world declined from 39 percent in 2000 to 32 percent in 2010. But the absolute numbers of slum dwellers have grown along with expanding populations. In sub-Saharan Africa, more than 60 percent of the urban population lives in slums—double the rate in Asian developing countries and much higher than the 24 percent in Latin America. Poor households typically spend a disproportionate share of their incomes on energy, so providing more energy-efficient housing can be a tool in the fight against poverty. But poor households will need grants and subsidies to help them weatherize or otherwise upgrade their homes.[47]

The first LEED Platinum mixed-use multi-family building in Southern California.

In principle, labor-intensive programs to improve the social and environmental aspects of housing and urban infrastructure could provide large numbers of green jobs—through new construction of buildings and retrofitting of existing ones and through production of insulation materials and efficient building components like windows, heating and cooling units, or appliances. Studies in a range of countries confirm that there is ample opportunity for greening existing construction work and generating additional employment and that more jobs are created than are lost in the energy-intensive industries that produce inputs like cement.[48]

Recent years have seen a degree of progress in greening buildings, though it is difficult to arrive at any worldwide figures. Although standards such as the LEED program in the United States have been replicated in a number of countries, there is no agreed worldwide definition of what constitutes green buildings. Also, allowance has to be made for a wide variety of climatic and other circumstances that require differentiated sets of standards. In the United States, it is estimated that 10–12 percent of new commercial construction and 6–10 percent of new residential construction

is green—figures that indicate an enormous potential remains to be tapped.[49]

Regulations and public policies can push the greening of buildings along. They include measures such as building codes, green procurement programs, appliance standards, energy- and water-efficiency requirements, mandatory audits, and the like. (See Chapter 10 for more on policies.)

In the European Union, the Energy Performance of Buildings Directive requires energy performance certificates to be presented to customers for building sales or leases. The European Commission thinks that by 2020, some 280,000–450,000 new jobs might be created, chiefly among energy auditors and certifiers, among inspectors of heating and air-conditioning systems, in the construction sector, and in industries that produce materials components and products needed to improve the performance of buildings. The insulation industry umbrella group Eurima provides more optimistic projections, estimating additional employment figures ranging from 274,000 to 856,000 jobs. And a study by the European Trade Union Congress and others estimated that up to 2.59 million jobs could be created by 2030.[50]

Some of the stimulus funds that were passed in several countries to address the economic crisis have been directed toward green building purposes. It has been estimated that this sector's 13 percent share of Germany's stimulus package of more than $100 billion will create some 25,000 manufacturing and construction jobs related to building retrofits. This builds on an earlier success story in Germany, where public funds for apartment and building retrofits triggered substantial additional private spending equivalent to $26 billion. By 2008, some 280,000 units had been renovated and about 221,000 jobs either were created or were saved from elimination—at a time when the construction industry faced a recession and the prospect of widespread layoffs. The same could happen in the United States, where the Better Buildings Initiative could result in the creation of 114,000 jobs.[51]

Greening the building sector requires adequately trained workers and professionals, such as architects. Denmark, Brussels in Belgium, Singapore, and Thailand are among the governments that have developed training programs. Many developing countries still fall short of the necessary expertise. In India, for example, more than 80 percent of the construction sector workforce is unskilled workers.[52]

Recycling. At the base of the brown economy is the large-scale extraction of natural resources. Mining of ores and minerals grew a staggering 27-fold during the twentieth century, outstripping the rate of economic growth. Now that easily exploited deposits have largely been exhausted, environmental impacts of mining are bound to worsen. Already, about three times more rock and other material needs to be removed now than a century ago in order to extract the same quantity of ore. A throwaway economy means that waste streams keep expanding along with mining. Worldwide, about 11 billion tons of solid waste were collected in 2010 (and an even larger, but unknown, quantity generated).[53]

All too often, waste management translates into landfilling, incineration, and shipment to other countries, either legally or illicitly. These practices impose an environmental and health toll on adjacent communities. By contrast, recycling, reuse, and remanufacturing of products permit a reduction in logging and mining; they save substantial amounts of energy and water by replacing the processing of virgin materials with greater reliance on scrap materials; and they avoid air, water, and land contamination associated with waste disposal. More than 1 billion tons of metals, paper, rubber, plastics, glass, and other materials are recycled each year. But that is only one tenth the amount of waste collected.[54]

Recycling is also good from an employment perspective. On a per-ton basis, sorting and processing of recyclables sustains 10 times as many jobs as landfilling or incineration do, and the manufacturing of new products from recycled materials or equipment employs even more people than sorting recyclables does. In industrial countries, recycling is a formal industry, often with a high degree of automation. In the United States, direct and indirect recycling employment runs to an estimated 1.4 million, and in the European Union, about 1.6 million.[55]

In developing countries, much greater quantities of recyclable materials are recovered by informal waste pickers than by formal waste management companies. Urban areas in these countries often have inadequate waste collection or none at all. Wastes typically end up strewn in streets, fields, and streams, as well as in open dumps. Many of the people engaged in waste picking and recycling in these countries are part of the informal economy.[56]

People who sift through uncontrolled dumpsites confront hazardous work conditions: they are exposed to a range of toxins and are vulnerable to intestinal, parasitic, and skin diseases. Earnings are often low and unstable. Moreover, municipal governments all too often regard waste pickers as expendable nuisances, frequently either ignoring them in policymaking or even harassing and persecuting them.[57]

An often-cited estimate puts the number of informal waste pickers at 1 percent of the urban population in developing countries. In absolute terms, a figure of 15 million people is sometimes mentioned in the literature. Mathematically, 1 percent today translates into a number as high as 26 million people. These numbers, however, are little more than educated guesses.[58]

Forming local and national cooperatives, waste pickers are becoming more organized in fighting for legalization, improvements in their social status, and better bargaining positions vis-à-vis municipalities and powerful intermediaries. Brazil has the most advanced group. The Movimento Nacional dos Catadores de Materiais Recicláveis emerged from years of local organizing efforts that had their origins in Porto Alegre and São Paulo in the 1980s. During the past decade, national legislation offered growing support. Waste picking has been recognized as a legitimate occupation. In 2010, the National Policy of Solid Waste mandated that informal recyclers be included in municipal recycling programs. The comprehensive national poverty alleviation plan (Brasil Sem Miséria) launched in June 2011 offers training and infrastructure support to waste pickers, and aims to achieve their socioeconomic inclusion in 260 municipalities.[59]

In various parts of the world, the last two decades have seen growing legal recognition of waste pickers as attitudes gradually change, strengthening of their organizations, integration into municipal waste management systems, and social inclusion. This has resulted in improvements in earnings and secured some social benefits. But Chris Bonner of Women in Informal Employment: Globalizing and Organizing cautions that "gains made by workers in the informal economy are often impermanent. There is a constant struggle not only to improve their situation, but often to merely hold on to what they have won."[60]

The global economic crisis is affecting the demand and market price for recyclables and compelling more people to rely on waste picking in the face of a lack of formal-economy jobs. Among the challenges this brings are moves toward waste management privatization in ways that sideline the pickers and their organizations and the emergence of new waste streams—particularly e-waste—that expose waste pickers to new occupational and health risks and will require a greater degree of training (to understand how to safely dismantle electric and electronic waste products, for instance) as well as proper equipment.[61]

Promoting Green Jobs Globally

To improve knowledge of green jobs trends and developments, governments need to craft detailed definitions and sector-by-sector criteria (as the U.S. Bureau of Labor Statistics is currently doing). Internationally, it would make sense to establish green jobs standards and certifications so that national data are comparable. Industry surveys or input-output modeling (as the German environment ministry has done in the renewable energy sector for several years) can help generate regular annual data. Green jobs data need to be integrated into regular national economic statistics.

Skills shortages could hamper the emergence of a green economy. To avoid this, governments should support a range of training efforts. A national skills mapping exercise could be undertaken with the goal of establishing green skill profiles in each industry, identifying strengths and gaps in the existing skills base, and creating a plan for overcoming gaps (as the regional government of Navarra, Spain, has done). Governments can also set up or facilitate the creation of green training centers and can encourage private companies and educational institutions to incorporate green jobs skills into courses, apprenticeships, and other workplace training. They should ensure gender balance and access by disadvantaged communities.

Green jobs are not necessarily or automatically "decent" jobs. Effective social dialogue between employers and workers, including collective bargaining arrangements, and broader public-private partnerships aiming for equitable outcomes are essential for decent work standards and social inclusion. Government action may be needed to establish and enforce decent wage standards and occupational health and safety rules. Governments may also need to pass social-inclusion legislation (as Brazil has done with regard to informal waste pickers).

To date, the emergence of green jobs has not come at the direct expense of jobs in polluting industries. Eventually, however, transitioning to a green economy does imply that such industries will shrink and perhaps disappear entirely. Governments should proactively create and fund "fair transition" programs for affected workers and communities, offering retraining and, if necessary, relocation assistance so people have an opportunity to find new livelihoods in the emerging green economy.

The nature of green jobs will vary according to economic sector and even to some extent country by country. Thus the specifics of the green jobs experience will naturally vary to some extent as well. Nonetheless, in order to facilitate the spread of green technologies and methods, it is important to share lessons—policy innovations and green roadmaps that have proved successful—as widely as possible. The United Nations can play a useful role in this context by establishing a UN Green Jobs Best Practices Unit (with inputs from UNEP and the ILO). Further a UN Green Jobs Coordinating Group could ensure policy cohesion among various agencies. An advisory council drawn from experts and stakeholders from business, labor, and civil society could help guide this work and analyze key developments, opportunities, and challenges.

A New Global Solidarity

A new global solidarity for sustainability must take root, ensuring that no one—no country, no community, no individual—is left behind. Unlike the conventional pattern of economic competition that produces—and indeed is expected to produce—winners and losers, the quest for a green economy needs to focus on win-win outcomes that render economic activities sustainable everywhere. There is already intense competition among manufacturers of green technologies and products, such as wind and solar energy, and government policies that reek of green mercantilism and protectionism. (See Box 1–2.)[62]

It is essential that cooperative models be developed for shared green development. A simple slogan therefore would be "avoid losers." Given shared environmental vulnerabilities on a small and increasingly crowded planet whose resources are being maxed out, there needs to be recognition that the winners will lose if the losers don't win.

For the rich of this Earth, greening action looks of necessity different than it does for those who aspire to greater wealth and for those who contend with poverty. In relative terms, the poor have to win more in a green economy than the rich do, so as to reduce and eventually overcome the stark differences in claims to the planet's remaining resources. Environmental sustainability is ultimately impossible without social equity. This requires that the rich reduce their draw on materials and goods in absolute terms.

Both environmental and social conditions have reached a state that requires a clean break with business-as-usual solutions. A key need is a rebalancing of public and private actions. Since the first Rio conference, in 1992, too much time and effort has gone into making market forces propel the greening of the economy. Market forces only work when they are properly regulated. Otherwise they tend

Box 1–2. Renewable Energy and Trade Disputes

U.S.-China Wind Subsidies. In September 2010, the United Steelworkers petitioned the Obama administration, asserting that the Chinese government provided millions of dollars in illegal subsidies to domestic turbine manufacturers that agreed to use key components made in China rather than imported parts. The union claimed this amounted to an unfair advantage and undermined U.S. companies' competitiveness in the Chinese market. The U.S. administration agreed to investigate the case and subsequently filed an official complaint with the World Trade Organization (WTO). After consultations, China in June 2011 agreed to halt its wind power subsidy program. Critics, however, argued that the steelworkers should push their own government to pursue more ambitious strategies, including adoption of a national renewable energy target. U.S.-China trade disputes could hinder future development of renewable energy technologies. The trade disagreement could also have been used to kick off a discussion on the need for WTO to legalize and regulate subsidies for alternative energy.

U.S.-China Solar Trade. In October 2011, seven U.S. solar panel manufacturers filed a complaint against the Chinese solar energy industry, accusing it of receiving illegal government subsidies and dumping completed panels in the United States under their marginal cost. The filing at the Department of Commerce and the International Trade Commission called for the U.S. government to impose high tariffs—more than 100 percent of the wholesale import price—on Chinese solar panels. In the first eight months of 2011, China exported $1.6 billion worth of solar panels to the United States. The Chinese Development Bank provided $30 billion in low-interest loans to solar manufacturers in 2010 alone, helping China

to claim the title of leading solar exporter. This helped push wholesale solar panel prices down from $3.30 per watt of capacity in 2008 to $1.20 in October 2011—a key factor in the much-discussed bankruptcy of U.S. manufacturer Solyndra. Chinese solar panel makers may move some of their operations to the United States in an effort to evade protectionist measures. The imposition of tariffs could also trigger Chinese retaliation: instead of purchasing raw materials for solar panel production from the United States, China could import them from German suppliers. Chinese officials claim that the steep tariffs would hamper the cooperative development of solar energy and undermine global support for clean energy.

Japan-Ontario FIT Dispute. In September 2010, Japan filed a complaint with WTO against Ontario's 2009 Feed-In Tariff (FIT), which offers renewable energy manufacturers a higher rate than conventional electricity suppliers receive for a 20-year period. The FIT is coupled with a domestic content requirement of 50 percent in 2010 and 60 percent in 2011. It has created 13,000 jobs and attracted $20 billion in private-sector investment so far. Japanese companies not meeting the domestic content rule argue it is discriminatory and that FIT encourages import substitution subsidies that are illegal under WTO rules. The FIT has come under scrutiny from the North American Free Trade Agreement, and the European Union joined Japan's complaint, claiming FIT is in "clear breach of the WTO rules." The irony is that Japan passed its own FIT legislation in August 2011, a policy driven in part by the Japanese government's decision to reduce reliance on nuclear power in the wake of the Fukushima disaster.

— *Miki Kobayashi*
Source: See endnote 62.

toward excess, create "externalities," and disregard social equity. The last 20 years have witnessed a certain abdication of public policymaking responsibility. It is time to rediscover this obligation. There is a need to recognize that "harnessing" the market requires more public policy, not less.

The policy suggestions that follow are not meant to be complete but rather suggestive—indicating the types of approaches that could help humanity achieve sustainability with equity.

A Network of Cooperative Green Innovation Centers. In order to spread green innovation as widely as possible, cooperative models are needed for green R&D and technology deployment. The *World Economic and Social Survey 2011*, for instance, refers to the successful experience of the Consultative Group on International Agricultural Research as an example of how to promote the rapid worldwide diffusion of new technologies via a network of publicly supported research institutions. This model could be adapted, and the *Survey* suggests that an international regime allow for "special and differential access to new technology based on the level of development" and that intellectual property rights be changed to accommodate the rapid diffusion of green innovation ideas.[63]

Global Top Runner. One way to harness market forces for sustainability is through an approach Japan has taken with its Top Runner program, which was established in 1998 and has helped to make its economy one of the world's most efficient. The program sets efficiency standards for a range of products that collectively account for more than 70 percent of residential electricity use. On a regular basis, products available in a given category are tested by advisory committees with members from academia, industry, consumers, local governments, and mass media to determine the most efficient model. That then becomes the new baseline for all manufacturers, driving a process

of continuous innovation and improvements. Adopting such an approach on a global level could promote leapfrogging for sustainability. This could have even more fascinating impacts if paired with a social top runner policy that counters a global race to the bottom of cheap wages.[64]

Green Financing. Inefficient products all too often have the advantage of seeming cheap. Green products can be difficult to afford when they have high upfront costs (even though they save consumers money over the product's lifetime). Reducing or eliminating this disadvantage is a key task in facilitating the transition to a green economy. This could be accomplished with the help of a public green financing program that offers preferential interest rates and loan terms for green products. Green financing would be even more effective if it were linked to a Top Runner approach— if the most efficient models also had the most attractive loan terms.

Durability, Repairability, Upgradability. Tax and subsidy policies do not differentiate products according to how well they are made. In fact, orthodox economics assumes that a product that does not last is preferable because it requires faster replacement and thus helps lead to greater economic activity. In a green economy, tax and subsidy policies should give preferential treatment to products that are durable, repairable, and upgradable.

Energy and Materials Productivity. Similarly, tax and subsidy policies, as well as other tools of public policy, could be structured to accord preference to companies that excel in improving the energy and materials productivity of their operations. This could be done somewhat like the Top Runner approach by setting standards in each manufacturing sector and evaluating performance on a regular basis.

Pricing for Sustainable Well-being. In the existing economy, consumers who buy larger quantities of a given product are often rewarded with price discounts, which encourages consumption irrespective of need. In a green economy, a reverse system of pricing should be introduced. It would allow consumption of goods in quantities that are consonant with the satisfaction of basic needs and a decent life at low, affordable prices. But usage beyond a certain threshold would only be possible at steeply rising prices per unit, in order to discourage overconsumption. In different countries, the precise definition of such thresholds would naturally vary. Dakar in Senegal and Durban in South Africa have adopted very low tariffs for an initial amount of water consumption. The price for water usage above that level rises steeply. Such a tiered pricing system should be adapted for a broad array of products and services.[65]

Reduced Work Hours. Today most people end up working long hours in an effort to earn enough to move with the crest of a never-ending consumption wave. Decent wages make this an easier process than if people feel they have to resort to debt. An economy and population that are less in thrall to consumerism might entertain an approach that seeks to translate increased productivity in the economy into reduced work hours rather than more consumption. Rich countries will need to undertake this transformation if they are to reduce their overall claim on the planet's resources and open up much-needed material and ecological space for the world's poor.

Economic Democracy. A large number of countries are run by at least nominally democratic processes, but there is no democracy in the economic sphere that determines so much of human life—the bulk of people's waking hours, their incomes, their careers and sense of self-worth. In the United States, for example, corporations now have the same free speech rights as people, yet the vast majority of people have no control over corporations that often bestride the globe and trump the democratic process by dint of having become "too big to fail." Companies that are bound more closely to the needs and interests of their own workforces and the communities they serve might play a more constructive role in creating a sustainable economy—less single-mindedly pursuing growth and profits at the expense of people and nature. There is limited experience with alternative, more participatory forms of running companies, such as the Mondragón Corporación Cooperativa (MCC) in the Basque region of Spain. While limits to corporate growth are likely a necessary element of a more sustainable economy, it does not mean that companies need to be local only. Worker-owned MCC is Spain's seventh largest company—with over 100,000 workers, annual sales of $20 billion, and 65 plants overseas. One key to a different type of corporation is greater participation by stakeholders and less influence by shareholders. (See also Chapter 7.)[66]

Transformational policies are needed if sustainable prosperity for all—present and future generations—is the goal. The alternative is a planetary triage that, to use the terminology popularized by the Occupy movement, may work for the 1 percent but not for the 99 percent. Policies need to reach far beyond technical fixes, limited changes in tax and subsidy policies, or other marginal efforts. The nature and rationale of the economic system will need to change in fundamental ways. From growing the economy at all costs, the central focus instead becomes an economy that permits ecological restoration and enables human well-being without materialism.

CHAPTER 2

The Path to Degrowth in Overdeveloped Countries

Erik Assadourian

In 2010, the Second Conference on Economic Degrowth for Ecological Sustainability and Social Equity in Barcelona, Spain, convened more than 500 participants from over 40 countries to discuss how to intentionally "degrow" the global economy. (See Box 2–1 for the definition of degrowth.) A variety of academic papers were discussed—from the mechanics of economic degrowth to strategies on how to pursue and communicate this challenging concept.[1]

The conference even drew attention to some radical (albeit unsanctioned) approaches to building the movement. At the peak of the global financial bubble, for example, Enric Duran—claiming to be an entrepreneur starting a new technology business in Spain—approached a number of banks to seek loans. He then promptly donated most of the 500,000 euros he collected to the degrowth movement (minus interest and taxes paid). Called by some a modern-day Robin Hood, Duran used the loose lending practices of the bubble era to engage in this act of what he called "financial disobedience" and help reveal the risks of a poorly regulated financial system while simultaneously generating resources to

help fund alternatives to the current unsustainable economic system. While undoubtedly unconventional, Duran's actions and subsequent arrest certainly drew attention to the movement.[2]

Degrowth in a globalized culture where growth is seen to be essential for economic success and societal well-being seems to be a political non-starter even for those who may be sympathetic. For most people, who deeply believe growth is essential to modern economies, it seems to be a recipe for economic and societal collapse. But the rapidly warming Earth and other declines in ecosystem services reveal that economic degrowth is essential and will need to be pursued as quickly as possible in order to stabilize Earth's climate and prevent irreparable harm to the planet and, in the process, human civilization.[3]

Already, the conversation is changing in the media and among scientists. The hope of preventing a temperature rise of 2 degrees Celsius is weakening. Numerous studies have found that humanity is now on a path to increase the average global temperature by 4 degrees Celsius. Most recently, the journal *Philosophical Transactions of the Royal Society*

Erik Assadourian is a senior fellow at the Worldwatch Institute and director of its Transforming Cultures Project. He is co-director of *State of the World 2012.*

Box 2–1. Defining Degrowth

Degrowth is the intentional redirection of economies away from the perpetual pursuit of growth. For economies beyond the limits of their ecosystems, this includes a planned and controlled contraction to get back in line with planetary boundaries, with the eventual creation of a steady-state economic system that is in balance with Earth's limits.

Degrowth should not be confused with economic decline. As Serge Latouche, a leading thinker on degrowth, explains, "The movement for a 'degrowth society' is radically different from the recession that is widespread today. Degrowth does not mean the decay or suffering often imagined by those new to this concept. Instead, degrowth can be compared to a healthy diet voluntarily undertaken to improve a person's well-being, while negative economic growth can be compared to starvation."

Ultimately degrowth is a process, not the end point. As Latouche notes, the end point is abandoning faith in the promise of growth as driver of development. Economist Tim Jackson puts this idea in a user-friendly way, calling for "prosperity without growth." However, that prosperity should not be confused with what is deemed prosperity by many today—a consumer lifestyle—as that depends on a growth economic model and overuse of Earth's natural capital. Instead, as Latouche explains, a prosperous society is one "in which we can live better lives whilst working less and consuming less."

Thus degrowth will be a step toward a more secure, sustainable, sane, and just future, helping to reduce the number and size of ecologically destructive industries and to reorient economies in ways that improve well-being, strengthen community resilience, and restore Earth's systems—a path that from any sane perspective would be hard to confuse with economic decline.

Source: See endnote 1.

even examined projections of a 4 degree increase not by 2100 but by 2060, following the path of emissions that society is currently on. This path translates to catastrophe for human society: massive shifts in population as coasts flood, areas hit by extreme weather and droughts, and diseases spread to new areas. And the 2011 climate talks in Durban did nothing to stop the world's rush to this future.[4]

With governments like Canada pulling out of the Kyoto Protocol and with a new climate agreement probably stalled until 2020, the world is in all likelihood in for massive ecological shifts, which needless to say are incompatible with a growing global economy. Indeed, in 2007 the *Stern Review on the Economics of Climate Change* projected that climate change could reduce global economic well-being anywhere from 5 to 20 percent (measured in per capita consumption terms), depending on how much human activities warm the world.[5]

These ecological changes are brought ever closer and made ever larger by people's continued belief that growth by all on an overtaxed planet is a useful pursuit. In the past half-century, growth has been understood as the cure-all to societal problems. In reality, while it may help sometimes, continued economic growth is at the root of ecological shifts that will cause far worse problems. As the Prince of Wales noted in May 2011, "Our myopic determination to ignore the facts and to continue with business as usual is, I fear, creating the risk of a crash which will be far more dramatic, and far harder to recover from, than anything we have experienced over the past few years."[6]

And while that may be evident to those who study environmental trends, society is so

committed to growth that even many environmentalists and sustainable development
experts still advocate for "green growth," or just
the decoupling of growth from material consumption. As Harald Welzer, author of *Mental Infrastructures: How Growth Entered the
World and Our Souls*, notes, "The current
debate on decoupling…serves above all to
maintain the illusion that we can make a sufficient number of minor adjustments in order to
reduce the negative environmental consequences of economic growth while leaving our
present system intact." But humanity needs to
radically transform the global economy, reducing its size by at least one third—based on the
conservative ecological footprint indicator,
which finds that humanity is currently using the
ecological capacity of 1.5 Earths—even while
the poorest one third of humanity needs to
increase total consumption considerably in
order to achieve a decent quality of life.[7]

The Curse of Overdevelopment

Ultimately, overdeveloped countries (and
overdeveloped populations within developing
countries) will need to either proactively pursue a degrowth path or continue down the broken path of growth until coasts flood,
farmlands dry up, and other massive ecological changes force them away from growth into
a mad dash for societal survival. If overdeveloped populations keep ignoring the looming
changes—keeping their proverbial heads buried
in the sand—then this transition will be bru-
tal and painful. But if a strategy of degrowth,
economic diversification, and support for the
informal economy is pursued now, before most
of societal energy and capital is focused on
reacting to ecological shifts, these overdeveloped populations may discover a series of benefits to their own welfare, to their long-term
security, and to Earth's well-being.

It is no surprise that overdeveloped countries also suffer from a series of ailments connected to overconsumption—since affluence
and development decoupled long ago for many
in these countries. The clearest indicator is
the obesity epidemic now plaguing most industrial countries and developing-world elites. In
the United States, two of every three adults are
now overweight or obese, reducing their quality of life, shortening life spans, and costing the
country an extra $270 billion a year in med-
ical costs and lost productivity due to early
deaths and disabilities. This epidemic may even
lead to the next generation living fewer years
than their parents did, primarily due to obesity-
related problems like heart disease, diabetes,
and certain cancers. Tragic statistics, but there
are many who prosper from this type of
growth: agribusiness, processed-food manufacturers, marketers, hospitals, pharmaceutical
companies, and others all profit from maintaining the status quo. The diet industry alone
earns up to $100 billion a year on obesity in
the United States. And the United States is not
exceptional on this front, merely a trendsetter.
In 2010, 1.9 billion people were overweight
or obese worldwide, up 38 percent over 2002,
even though total population rose 11 percent
in that time.[8]

Obesity, unfortunately, is not the only side
effect of overdevelopment. Increased debt burdens, long working hours, pharmaceutical
dependence, time trapped in traffic, even
increased levels of social isolation stem at least
in part from high-consumption lifestyles.
Indeed, while many modern advances—personal transport, single-family homes, televisions, computers, and electronic gadgets—seem
to have improved human well-being, in reality
these advances may have imposed significant
sacrifices on consumer populations without
their knowledge or consent.[9]

More broadly, along with reducing the
physical and societal side effects of the obsessive pursuit of growth, pursuing degrowth
would reduce the ecological impacts of the
human economy, as some populations would

consume less food, resources, and energy. Perhaps the most important but least tangible outcome of this would be to reduce the loss of Earth's resiliency, which humanity and all species depend on completely for their ability to survive and thrive.

Of course, it is simple to advocate for the sanity of degrowing the ecologically destructive global economy. But when growth is one of the fundamental sacred myths of modern culture, and when economists, the media, and political leaders routinely wring their hands whenever the economy contracts, shifting paradigms 180 degrees will be extremely difficult. Instead, degrowth will need to be pursued very strategically—working simultaneously on a variety of complementary fronts.

Reducing Overall Consumption by Overconsumers

At the heart of degrowth will be dramatic shifts in individual and collective consumption patterns. A large percentage of people's ecological impact comes from food, housing, and transportation. These sectors will need to be dramatically overhauled so that people in overdeveloped countries choose to live more simply, in smaller homes, in walkable neighborhoods, traveling less by car and plane and more by foot, bicycle, and public transit, and eating less and lower on the food chain. Moreover, people will need to own less "stuff"—from electronics to appliances, from books to toys—that requires massive amounts of resources and produces considerable waste. Indeed, when adding up all indirect and direct forms of consumption, in 2000 the average American used 88 kilograms of resources a day and the average European 43 kilograms a day—numbers that need to contract tremendously to be sustainable, especially in the context of growing consumption demands by developing countries.[10]

This presents a formidable challenge, as

growth and consumerism are celebrated by an advertising industry that spent $464 billion worldwide in 2011 marketing the consumer lifestyle, by Hollywood and the global film industry, and by the media more broadly. A few cracks are appearing, however, in what were once solid traditions of the growth-centric consumer culture. Some American teenagers, for instance, are no longer rushing to get their driver's licenses—previously an essential rite of passage to adulthood. In 1978, half of the 16 year olds in the country got their license; by 2008 the number had fallen to 31 percent. Even by age 19, while 92 percent of teens had a license in 1978, only 78 percent had one in 2008. And this is a trend that now seems to have persisted even beyond teenage years: the percentage of total miles that are driven by people in their twenties fell from 21 percent in 1995 to 14 percent in 2009. Between the expense of cars and gasoline, traffic, rising environmental awareness, and shifts in technologies—with teens usually now connected online with friends—young people are finding less need for cars and more barriers to using them. Of course, this shift brings problems of its own, with the average U.S. teen now spending eight hours a day consuming media, but it does reveal that even long-standing traditions can become much less relevant over time.[11]

These shifts in deeply rooted consumption patterns will have to be replicated hundreds of times over in dozens of sectors—food, housing, transport, electronics, travel, pets, clothing, appliances and so on. And with changes so extensive, few individuals will be willing to make what they see as sacrifices—even if the products' downsides are made clear. (See Box 2–2.) Cultures quickly normalize certain goods, shifts in infrastructure often require them, social networks reinforce use of these goods ("keeping up with the Joneses"), and it is psychologically easy to convert a luxury item into a perceived necessity. Today, more than half of Americans view air conditioning and

Box 2–2. Sacrifice and a New Politics of Sustainability

Many commentators who argue that a sustainable society requires profound change also believe that this would involve considerable sacrifice in wealthy consumer societies. And that, they pessimistically assert, is just not going to happen: most people are too self-satisfied, apathetic, or uninformed to sacrifice willingly. But in fact sacrifice is a familiar part of everyday life and can be consistent with an inclusive sense of self-interest—although it can also be foisted on people unjustly.

A person can willingly sacrifice, giving up one thing of value for something more valuable, such as consuming less to save for a child's education. A person can also be sacrificed, as when a poor community bears the health effects of a toxic incinerator. This vital distinction about sacrifice is often overlooked—and is shaped by people's views of justice and effectiveness.

Recognizing the sacrifices that people already make can foster a more balanced consideration of political and policy choices. Rather than seeing the task as convincing people to sacrifice, it is possible to establish a dialogue about how certain luxuries or conveniences might be traded for gains in quality of life for all. The point is to neither call for sacrifice nor avoid talk of it, but to broaden the conversation about choices and challenges.

When those calling for sacrifice do not follow suit, those being called to sacrifice may perceive themselves as victims, rather than agents, and resist calls to sacrifice. When U.S. politicians push for emissions reductions in China and India, where per capita emissions remain radically lower, as a precondition for American action, it has the character of an unfair distribution of burden—of calling for others to sacrifice, rather than shared sacrifice. Sharing the

burden, and clearly acknowledging that these others are already giving up something of value, can go a long way to countering this hypocrisy and paternalism.

Sacrifice begets anxiety when people are afraid that what they give up will be wasted. To sacrifice willingly, this anxiety must be tempered with the hope that what is given up will lead to future good. But this hope can rarely be sustained through individual action alone, because the likelihood of success is diminished by collective action problems. A person might think, "If I act when others don't, I'll incur costs without social benefit; if I don't act when others do, I'll share in the benefit without cost." By contrast, when action is coordinated, new opportunities become feasible: large-scale investment in infrastructure and renewable energy, land use and urban planning to foster walkability and reduce car dependence, incentives for "green" jobs.

Such actions are not painless: public investment requires taxes; land use policy generates winners and losers; green jobs may be at the expense of "brown" ones. Yet such measures can reduce coerced and inequitable sacrifice now and temper the coerced and inequitable impact of climate change and other environmental harms in the future.

To rethink sacrifice is not to offer a specific set of policies. It is a way of thinking and talking about the challenges of sustainability that opens a political dialogue at precisely the point where it is often shut down. People must build on the radical hope that the future can be a better one for which it is worth taking action, even if that action comes with certain sacrifices. In a world with no guarantees, it is this hope that can inspire change.

—John M. Meyer
Humboldt State University
Source: See endnote 12.

clothes dryers as necessities, while new products like smartphones and high-speed Internet are also becoming quickly perceived the same way. Thus to reduce overall consumption, just encouraging people to change their behavior will be far from sufficient. Rather, government and business will need to play a central role in editing consumers' choices.[12]

"Choice editing," at its simplest, is exactly as it sounds—editing people's choices toward a certain end. Unfortunately, for the past 50 years that end was to stimulate economic growth and consumption. But the same strategies can be applied to promote degrowth and sustainability. Shifting the billions in government subsidies toward healthy sustainable goods—such as providing subsidies for small-scale organic farms rather than giant commodity producers or shifting tax credits from homeownership to living in small, efficient homes that are owned or rented—could make consumption patterns much more sustainable. Of course, choice editing takes finesse: total bans on some goods can lead to hoarding and political reactionism. But even subtle taxes significantly shift consumption behaviors. When Washington, DC, added a 5¢ tax on plastic bags in January 2010, use of these bags plummeted from 22.5 million to 3 million—in one month. And the $2 million in annual revenue collected from the tax is being used to help clean up the tons of consumer refuse polluting the Anacostia River, a long-suffering waterway that flows through the nation's capital.[13]

Businesses, too, can play a role in choice editing, making it clear to customers what the healthiest and most sustainable choices are, such as by labeling products using health and sustainability criteria or by pricing healthier and more-sustainable products favorably. Walmart announced in early 2011 that it would lower the price of its produce, reducing costs to customers by up to $1 billion, and would work to lower the amount of salt, added sugars, and unhealthy fats in their packaged foods. These behind-the-scenes changes could do a lot to shift consumers' behaviors, leading them to eat more vegetables and fewer processed foods.[14]

While many companies are open to editing their product lines to be more sustainable—and are already doing so—few companies will be bold enough to encourage people to not buy their products at all, as their bottom line depends on total sales. But one company in September 2011 garnered considerable attention for doing just that. Patagonia, an outdoor clothing manufacturer, urged its customers to not buy its products unless they really need to. And even then, Patagonia encouraged potential customers to consider buying its products used instead, as "the environmental cost of everything we make is astonishing." The company set up a partnership with eBay to help customers resell used Patagonia products—a surprising move, as the company receives no share in those sales.[15]

Although Patagonia's primary motive is to prevent "environmental bankruptcy," which, as the company notes, is being driven by the consumer culture, there is business savvy present too. The marketing value of these efforts may more than make up for any lost sales, as they increase loyalty of their "green consumer" customer base. And there's a first-mover benefit too. As the company's forecasters are reading the economic and ecological tea leaves, they already must recognize that in coming decades more people will most likely buy less stuff and more products that will last—so developing that brand advantage now will lead to long-term returns for Patagonia, even in an overall contracting economy.[16]

Beyond choice editing, there are also many groups helping to change specific consumption patterns. Take burials. In the United States, 3 million liters of embalming fluid, 104,000 tons of steel, and 1.5 million tons of concrete are used each year to bury the dead. Burials there produce more than 1.5 million tons of carbon dioxide (CO_2) emissions and cost the average family about $10,000, in what is essentially a tax on the grieving. The good news is that there are efforts to shift these trends, burying people without chemicals in natural burial grounds that create new community parklands, which in turn create new space for biodiversity and help serve as new carbon sinks—a much better model than today's pesticide-sprayed, grass-covered cemetery. And these shifts in burial process are helping transform this essential human ritual so that it reminds mourning families of humanity's part in the broader cycle of life—replacing efforts to delay decay with a celebration that with this loss comes new life.[17]

Similarly, the Slow Food movement is working to shift dietary norms away from meat-centric, highly processed, ecologically destructive, unhealthy food back to enjoying the preparation, cooking, and eating of "good, clean, and fair food." As food is such an emotionally evocative topic, Slow Food has tapped enormous interest in how people eat, and today it has over 100,000 members in chapters in 132 countries.[18]

Even when a broader transformation of diet is beyond the reach of some—as not everyone has time to slow down when they are simply trying to make ends meet—there are gentler entry points for shifting food and other consumption patterns toward lower-impact norms. The Meatless Monday Campaign, for example, encourages individuals to forgo meat once a week as a way to reduce the significant health and ecological impacts of meat consumption. While this campaign was launched by The Johns Hopkins Bloomberg School of Public Health in 2003, the practice was actually first instituted by the U.S. government during World War I, and then again in World War II, in order to ration meat for the troops. During the first war, over 10 million American families and 425,000 food dealers pledged to go meatless on Mondays. Although the new campaign has not reached that level, it has spread to a variety of countries, including the United Kingdom, Belgium, Israel, and India. The French cafeteria management company Sodexo has also come aboard, spreading the initiative to the 2,000 corporate and government and the 900 hospital cafeterias it manages.[19]

Altering such deeply set cultural norms will take continual intervention at a number of levels by as many actors as possible. As an exhibit on the government's effect on the American diet noted, to shift Americans' diets during war time "the battle was fought with squadrons of celebrities, anthropologists, and cartoon characters, and a flotilla of films, radio programs, pledge drives, and posters." This deep level of intervention will once again be necessary to shift current consumption patterns.[20]

Distributing Tax Burdens More Equitably

Today, the gap between wealthiest and poorest has grown to dramatic proportions. (See Chapter 1.) While this is a social justice issue, it is also an environmental issue, for the more wealth someone possesses, the more that person consumes. Ultimately, on a planet with 7 billion people, an ecologically sustainable annual income is on the order of about $5,000 per person per year (in purchasing power parity terms)—far below the current understanding of western poverty levels. Beyond this level, individuals purchase larger homes, more appliances, air conditioning, electronic gadgets, even air travel.[21]

But how does society intentionally converge global incomes toward a lower norm? Shifting tax burdens will play a central role, as will redistribution of job hours—reducing the length of the average work week will free up work and income for others while also helping to reduce overall incomes of the overworked. In the process, there would be not only ecological and economic benefits but considerable social benefits as well. Research has shown that more-equitable societies have less violent crime, higher literacy levels, are healthier and less over-weight, and have lower teen preg-nancy and incarceration rates.[22]

Better distribution of incomes has a clear impact on human develop-ment, as the *Human Development Report 2011* reinforced. This report by the U.N. Development Pro-gramme found that when inequali-ties in income, health, and education are taken into account, several of the wealthiest nations fall dramatically in human development rankings. The United States falls from fourth to twentieth in the rankings, for example, while countries with high equity fare better: Sweden goes from tenth to fifth and Denmark increases from sixteenth to twelfth.[23]

One of the most direct routes to shifting taxes is simply adjusting income tax burdens. This may sound politically impossible in coun-tries like the United States—where anti-tax political movements like the Tea Party exist. But efforts like Occupy Wall Street may open up new political possibilities, especially if Americans start to recall their history. During World War II, marginal income tax rates on those earning more than $200,000 a year peaked at 94 per-cent. And while greater levels of influence by moneyed interests over the political system will make this more challenging now, there is no legal barrier that would prevent Americans from doing this again. Given that the threat of today's environmental crisis is even graver for U.S. national security than World War II was, this type of shift should be on the table. Activists and researchers could benefit from studying the messaging used to get these tax increases through Congress and try to apply those lessons to current tax reform efforts.[24]

Extremely high income tax rates are not necessarily the only (or the best) path to take

Poster of self-interested creatures encouraging people to go meatless on Mondays

© The Monday Campaign, Inc.

if other taxes are adjusted as well. Recently, much attention has been given to a small tax on financial transactions, which could both help make financial markets less volatile and generate revenue for sustainable development. While this has been advocated by some since first proposed by economist James Tobin in 1972, the idea suddenly has new life behind it. Occupy movement protestors have included this in their demands, and several influential individuals—including billionaires Bill Gates and George Soros—have publicly backed it, urging that the tax be used for development aid. The European Commission is now con-sidering a fee of $10 per $10,000 of financial transactions by 2015, which could generate $77 billion annually in new tax revenue. And while the idea has critics, the United Kingdom already imposes a $50 tax on $10,000 of stock trades, so the tax is clearly workable, both financially and politically.[25]

Ecological taxes could also be strength-ened and even be used to offset burdens on the individuals most affected by the contrac-tion of certain polluting industries and dis-ruptions brought about by degrowth. In late 2011, Australia passed a tax of $23.78 per ton of carbon, which is projected to cut CO_2

Billboard Liberation Front

 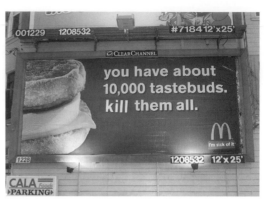

McDonald's billboard before and after being "jammed" by Billboard Liberation Front activists

emissions by an annual 160 million tons by 2020 while generating $15.5 billion a year by 2015. This is good news, since earlier in the year the Australian government announced it had to cut funding for environmental programs due to costly flooding—flooding that environmentalists connected to climate change. Governments will clearly need revenue both to prevent additional environmental disasters and to adapt to a warming, more disaster-prone, world.[26]

Finally, one other industry ripe for taxation is advertising. In the United States, corporations' advertising budgets are currently tax write-offs, but ending this and even modestly taxing these expenditures could yield significant new revenues. In 2011, advertising expenditures in this one country alone were $155 billion. Assuming the elimination of a tax write-off at a modest 20 percent corporate tax rate, that translates into $31 billion in new tax revenues. Add to that a tax on advertising for unhealthy and unsustainable products—such as junk food, fossil fuels, and automobiles—and this could provide new funds for marketing the elements necessary for normalizing a less-consumptive society; it could also deter the marketing of unsustainable and unhealthy products.[27]

What would all these new taxes be used for? First, not all taxes need to go into gov-

ernment programs; they can be redistributed in a way that increases societal equity and that compensates groups most affected by the shift to a degrowth economy—providing people with transitional support, key social services, and training in new skills. But simply rebuilding public infrastructures would use up a sizable percentage of the taxes collected. At the most basic, this includes improving public water and sanitation systems, accelerating the transition to the efficient use of renewable energy, and replacing car-centric infrastructure with one centered on bicycles and public transit. To these upgrades could also be added new community centers, swimming pools, hiking trails, and libraries that lend not just books and media, but games, toys, and tools—all of which will help convert what are increasingly private luxuries once again into public goods. In the process, these new developments could ease people's frustration with shrinking levels of wealth and diminished reserves of private goods by providing new opportunities to play, learn, and socialize.

The new funds can also be used to prepare for an unstable future. Governments have key roles to play in, for example, restoring ecosystems like forests and wetlands, supporting entrepreneurs to create new small, local farms, and actively preparing for the now inevitable changes that a warming world will bring (including in some cases abandoning certain areas altogether). The Netherlands is already proactively addressing climate change—not surprising, considering much of the country is just above or even below sea level. In *Hot: Living Through the Next Fifty Years on Earth*, Mark Hertsgaard describes the extensive lengths to which the Dutch government is going to prepare for a warming world—steps many people would find too extreme.[28]

The Dutch government has created a 200-year plan to adapt to climate change, spending $1 billion a year to implement it. Coastal hotels are being closed to make way for new protective dikes, and farms are being converted to lakes, with the long-term public interest taking precedence over short-term private interests (though owners are compensated when dislocated). These significant investments will need funding support an estimated $2–6 billion a year in the case of the Netherlands—as will simply adapting to surprise weather disasters. Twelve disasters in the United States in 2011 cost over $1 billion each—causing $52 billion of damage (more than the global total of disaster damage in 2009) and setting a new record for total number of devastating disasters to hit the country in one year. With growing instability from a changing climate, new tax revenue will be necessary to ensure that there is enough money in the coffers to respond to the next surprise that nature throws at humanity.[29]

Sharing Work Hours Better

Another way to improve access to incomes—one that may more easily get through conservative legislatures—is to do a better job distributing work hours. Since World War II, the 40-hour workweek has been seen in many western countries as "normal." Few countries have reduced working hours from this standard—even as technologies and productivity

have improved and as reductions could help shrink unemployment. More striking is that if the true average workweek were calculated—taking into account the unemployed, the underemployed, the part-time workers and full-time workers, and those working excessive hours—it would be much lower. The New Economics Foundation found that the average Briton worked 21 hours a week in 2010.[30]

A better distribution of job hours among all working-age individuals will not only help reduce poverty, it could significantly improve the quality of life for the many people working too much and it could reduce their ecological impacts. Psychologists Tim Kasser and Kirk Brown have found that longer working hours correlate negatively with life satisfaction levels and positively with ecological footprint. Moreover, if the work-hour reductions are actively supported by the right social marketing messages, more of people's time could be directed toward living more sustainably: bicycling instead of driving, drying clothes on clotheslines, cooking instead of buying packaged foods or going to restaurants, taking local "staycations" instead of exotic vacations, playing board games instead of going out for expensive entertainment, going to the library instead of the bookstore, gardening, volunteering, and taking care of children and elderly parents, all of which could help improve health, social connectedness, and community engagement—in other words, well-being.[31]

Although many people would be willing to earn and spend less, few have the opportunity to choose this, as businesses receive incentives to hire full-time employees. Some countries have already taken steps to remedy this. The Netherlands, for example, helps people cut their working hours to three-quarters time by requiring employers to maintain individuals' same hourly rate of pay and pro-rated benefits at the reduced working rates. During the recession, the German government helped businesses retain employees they might otherwise have laid off in a program called Kurzarbeit. The program, meaning "short-work," enabled companies to pay workers only for the hours worked while the government made up the difference (at up to two-thirds time). The program has supported 1.5 million workers across 63,000 companies, reducing layoffs by 300,000–400,000 and helping to keep the unemployment rate in Germany to a 17-year low. Through innovative programs such as these, governments can help save over-all costs and avoid societal disruptions caused by unemployment while also helping with the transition to shorter hours.[32]

Companies can also create more space for leave time—such as through more vacation time, more maternity and paternity leave, or opportunities to job share. Some advocacy groups like Right2Vacation.org are lobbying to get a minimum one-week vacation for all U.S. workers, as the country has no law requiring vacation, and half of all workers there get a week or less of annual vacation time. The more vacation time built into work schedules, the shorter the average work week will be and the more overall jobs will be available. The same is true with parental leave. The United States is one of only four countries in the world that does not have paid maternity leave. Providing generous maternity leave not only helps mothers bond with new infants and increases the probability of breastfeeding, it also reduces total hours worked across the population and thus helps distribute job hours more broadly. In Sweden, new parents receive a combined 480 days of parental leave, with 390 days sup-ported at 80 percent of their salary—which, not surprisingly, encourages many of these new parents to work less.[33]

Overall, there will need to be better distri-bution of job hours and in due course a con-traction of the consumer economy. But mostly the contraction—if managed—could be of goods and services that are artificially stimulated solely to make a profit and that cause both ill health and ecological degradation. Cigarettes, junk food, cars, weaponry, alcohol, cosmetics, disposable packaging, and many other sectors of the economy produce jobs, but should these oftentimes socially irresponsible industries be maintained at current levels just to sustain over-all employment levels? Or should society shift the economy to provide a healthy and sustain-able way of life along with work that does not undermine the planet and humanity's long-term well-being? Shrinking and even phasing out certain economic sectors and replacing them (when beneficial) with other economic pursuits will be an essential step in degrowth, even if to some it looks like "reverse progress."

Cultivating a Plenitude Economy

Sociologist Juliet Schor has spent decades study-ing work hours and the high levels of con-sumption that accompany working too much. She has drawn attention to these issues in pop-ular books like *The Overspent American* and *The Overworked American*. In 2010 she pub-lished *Plenitude*—a term that refers to great abundance or the condition of being bountiful. In this book, Schor calls for the controlled reduction of the consumer economy, with more people supporting themselves with a diverse set of formal and informal economic activities, including self-provisioning and trading of food and artisan goods as well as maintaining and repairing goods for longer periods. Ultimately, shifting some portion of the household econ-omy to these informal economic activities "expands a household's options with respect to employment choices, time use, and consump-tion," notes Schor. "The more self-provision-ing one can do, the less income one has to earn to reproduce a standard of living."[34]

The combination of intentional shifts in working hours and inevitable market contrac-tions could help accelerate this plenitude model. The recession in the United States has played a role in increasing the number of people living in multigenerational households to 51.4 million Americans, up 10 percent between 2007 and 2009. When different generations share a home, living costs can be reduced considerably—in housing, utilities, and transportation. This helped keep U.S. poverty rates of multigenerational households lower than those of other households, even as their median income levels were lower. In addition, elderly parents can help with children (and also be looked after if necessary), lowering both child and elderly care costs. And more household economic activities can be taken on—such as gardening or raising livestock. Although these are time-intensive activities, they can be done more easily when more people share the time burdens.[35]

U.S. goverment social marketing poster from 1917

Library of Congress

Multigenerational housing should be actively celebrated by popular culture and supported with government incentives, as it will significantly reduce ecological and economic costs while redeveloping social capital and neighborhood density. It may even open up new entrepreneurial opportunities: Lennar, a U.S. housing developer, has created a new line of multigenerational houses to sell to people embracing this demographic shift.[36]

Strategic social marketing could help this. Marketers have been targeting multigenerational households in the United States since the recession began, primarily to sell them more stuff. But if the government and public interest groups also reached out to these households—offering pamphlets, online videos, and workshops on canning, basic repair, sewing, and so on—this could help encourage diversification of household livelihoods and help normalize both this housing strategy and broader aspects of plenitude living.[37]

The contribution of this sector should not be underestimated. In the United States, during World War II, 40 percent of vegetables consumed by households were grown in home and community gardens. Gardening could reduce both household food costs and the ecological impacts of agriculture if people are taught food cultivation strategies that emphasize organic and integrated pest management methods. As climate change disrupts large scale agriculture and as food-insecure countries ban the export of grain, backyard and community gardens could play a substantial role in food security and community resiliency. Individual gardens have played an essential role in Cuba, for example, since the collapse of the Soviet Union reduced its access to cheap oil. In Havana alone, more than 26,000 food gardens are spread across 2,400 hectares of land, producing 25,000 tons of food annually.[38]

Juliet Schor is optimistic that, over time, people will get disillusioned with the 9-to-5 work/spend/consume lifestyle and more of them will actively seek out a plenitude lifestyle, working fewer hours at paying jobs and helping to rebuild local economies. Some civil society organizations are working to accelerate this transition in a variety of ways. For decades there have been efforts to encourage people to live more simply—working less, buying less, and enjoying the greater amount of time they have with friends, family, or hobbies. "Voluntary simplicity" initiatives have taken on many forms—from study circles and Simple Living television shows to annual "Buy Nothing" boycotts and websites that help people share and exchange unneeded goods. They have helped millions of individuals to "downshift" their spending.[39]

Outdoor oven at the Sirius Community, an ecovillage in Massachusetts

Many religions have been active in encouraging adherents to live more simply as well—a role their ancient teachings deeply support. From Pope Benedict XVI advocating for a less commercialized and more meaningful Christmas, to Jews developing a new Eco-Kosher standard to encourage more-sustainable food choices, to Muslims advocating for a Green Ramadan—with the annual fasting ritual being expanded to include eating more locally grown food and reducing carbon footprints by 25 percent—a wide variety of religious efforts are encouraging simpler living. In 2009, the Catholic Church created the St. Francis Pledge, named after the monk who lived an ascetic life in the thirteenth century and is the patron saint of the environment. People who take the pledge are asked to reflect on their eco-

logical impact, alter choices and behaviors to reduce their carbon footprint, and advocate for better care of God's creation. Initiatives in the religious community along these lines are still fairly small. But given that 80 percent of the people in the world identify themselves as religious, expanding the leadership role of religions could dramatically accelerate the transition to a plenitude society.[40]

In the United States, Common Security Clubs have started working over the past few years to proactively rebuild social capital and links in the informal economy. Community members come together in groups of 10–20 to assess how they could help each other, exchanging skills and resources—from tools and trucks to time and an extra room in someone's home. Neighbors are once again starting to do what neighbors used to do: help each other. Facilitated by the Institute of Policy Studies, these clubs are spreading around the country, growing especially in church communities or small towns where a basic level of social cohesion already exists. Along with fostering community resilience, these groups also teach people about broader economic issues and mobilize members to become politically active.[41]

On a larger scale, the Transition Towns movement founded in 2005 is working to reduce community energy usage and relocalize economies and food systems in order to make communities more resilient in the increasingly constrained future. There are now nearly 400 communities in 34 countries recognized as official Transition Town Initiatives. These towns have brought together diverse sectors of society to create community gardens, tool exchanges, and waste exchanges between businesses, for instance. There is even an eco-circus in Shaftesbury, England, that uses clowns, humor, and performances to teach children and their families about climate change and sustainable living.[42]

One U.S. effort—the Oberlin Project—is taking the Transition Town model to a new level, working to sustainably redevelop the region around Oberlin College in Ohio. By using the creative energy, talent, and financial resources of the college community, this project—if successful—could help drive progress toward the participants' deep green vision of a carbon-neutral city and a "20,000 acre greenbelt of farms and forests" that will form the backbone of a robust local economy. As David Orr, Oberlin professor and visionary behind the project, notes, the entire effort will also serve "as an educational laboratory relevant to virtually every discipline."[43]

Ecovillages also play an essential role in modeling the plenitude economy. With hundreds spread around the world, many of these communities have pioneered a sustainable and resilient way of life for decades—exploring the frontiers of permaculture, alternative building materials, renewable energy systems, even lost skills like horse-powered agriculture. The majority of these ecovillages also make it a central mission to spread these skills to broader society, regularly holding retreats and workshops for visitors from around the world.[44]

Ecovillages have also rediscovered traditional wisdom that will play an important role in a constrained future. At The Farm, in Tennessee, for instance, some midwives are a repository of knowledge about skills that had been lost as the U.S. medical system eagerly embraced modern technologies in delivering babies. Today, in the United States, one third of pregnant women have their babies by Cesarean section (C-section)—putting themselves and their babies at often unnecessary risk. Many of these procedures are due to misinformation, cultural pressures, even time pressures on hospital staffs. The Farm's midwifery program has helped train many new midwives and has drawn attention to the overmedical-

ization of childbirth, showing that Cesarean sections are rarely necessary. Of the 3,000 births handled by The Farm's program since 1971, under 2 percent have been C-sections. Considering the ecological and financial resources needed for surgery (along with the risks to mother and baby), the degrowth of unneeded medical interventions will be essential, and society will need to look to innovators like these for inspiration and advice on how best to treat medical needs sustainably and safely.[45]

Government can play a valuable role in cultivating a plenitude economy as well. As the consumer economy has come to dominate, many of the skills needed for a plenitude economy have been lost and will need to be relearned. Governments could support training—both directly and through funding nonprofit or community organizations—to help redevelop basic household skills. This is already happening in several European countries. More than 1,200 "social farms" have been established in France, for example, and over 700 in the Netherlands. These ventures use farming as a means of creating jobs and new skills, as well as offering opportunities to reconnect with nature, build community connections, and in some cases help rehabilitate mentally handicapped populations—not to mention provide sustainable and local sources of produce.[46]

Governments could also help individuals and communities get involved in the management and restoration of public and marginal lands. Although this certainly would not appeal to everyone, growing numbers of people seek opportunities to lead a more-traditional lifestyle. With support, a new, bolder version of "the back to the land" movement of the 1970s could take off. At the moment, the financial downturn in Greece has led to a growth in the agricultural sector of 32,000 jobs, even as unemployment shot from 12 to 18 percent. This agrarian way of life, with the right training, would not just be low impact but could be actively eco-restorative, if it is based on proactive, sustainable management of ecosystems.[47]

The Colombian village of Gaviotas demonstrates just how much can be achieved by a small community committed to ecological restoration. This village of 200 people was established on degraded savanna 30 years ago and since then has replanted over 8,000 hectares of surrounding land with forest—an area larger than Manhattan. This forest now provides the village with food and tradable forest products, while absorbing 144,000 tons of carbon a year as it grows. Supporting this type of community-driven ecological restoration—particularly in ways that encourage extremely low-consumption lifestyles—could help greatly in the pursuit of sustainability.[48]

The last important point about a plenitude economy is that it will free up ecological capacity for people who are living in true poverty and maintain the key services that society does not want to lose in a constrained future—hospitals, vaccines, antibiotics, basic education, energy production, clean water infrastructure, and so on.

Moving Toward Degrowth

Ultimately the idea of decoupling growth and prosperity is no longer a utopian dream but a financial and ecological necessity, as Tim Jackson puts it. Right now, however, prosperity is deeply understood as consuming ever more and growing ever larger. Thus, moving toward degrowth will involve redefining prosperity altogether—resurrecting traditional understandings of what this word means: health, social connectedness, freedom to pursue hobbies, and interesting work.[49]

Communicating this shifted meaning will be a challenge, especially with 1 percent of the global economic product spent each year marketing consumer goods and services and the romanticized idea that they will bring happiness. To succeed, effective communication strategies will need to be pursued across many realms—from the Internet and the classroom to the voting booth and the living room. Fortunately, some promising initiatives may point the way.[50]

First, "social marketing" strategies are being used to challenge overconsumption and even growth. The Story of Stuff project has been effective in challenging the use of cosmetics, electronics, bottled water, and even the spending of unlimited funds on political marketing. The New Economics Foundation also created a short film that captures the absurdity of infinite growth flawlessly, applying this goal to a hamster. As the film reveals, if a hamster did not stop growing as it reached adulthood, it would be 9 billion tons on its first birthday and "could eat all the corn produced annually worldwide in a single day. And still be hungry." As the narrator concludes, "There is a reason why in nature things grow in size only to a certain point, so why do economists and politicians think that the economy can grow forever?" Similarly, and seen by much larger audiences, popular Hollywood films like *Avatar* and *WALL·E* are also playing an important role in drawing attention to the possible devastating outcomes of a continued obsession with growth and consumerism—literally the destruction of planet Earth.[51]

Beyond film, there is now a degrowth movement, with annual conferences on this topic and a budding political movement; degrowth political parties exist in several countries, including France and Italy. A variety of publications and websites are devoted to the subject, including a monthly magazine in French, *La Décroissance*, and an Internet hub for the topic at DegrowthPedia.org. As degrowth is brought more openly into dialogues and as progressive politicians articulate positive visions of it, the concept can move from the realm of the taboo to the more normal, creating space for mainstream media and political parties to break away from assumptions that growth is always good.[52]

More aggressive efforts in the classroom and academic settings may help too. In 2009, the Adbusters Media Foundation—which is credited with starting Buy Nothing Day, Turnoff the TV week, and Occupy Wall Street—started a campaign to get economics students to challenge their professors to adapt the flawed neo-classic economic model to ecological realities of life on a finite planet. By putting up posters, starting debates, sending open letters, and even walking out of class—as a group of Harvard students did in November 2011—students hope Economics Department curricula will start teaching a "new economics—open, holistic, human-scale." In a similar but less confrontational manner, groups like Net Impact, which has 20,000 members across six continents, are working with Business School professors and administrators to integrate sustainability and social responsibility courses into academic curricula and to help find socially responsible business opportunities for graduates.[53]

Subtly harnessing popular culture icons to question growth can also play an important role. One example is a new eco-educational scenario for *The Settlers of Catan*, an award-winning board game with over 18 million copies in print in 30 languages. The scenario, *Catan: Oil Springs*, not only incorporates clear side effects of growth, such as pollution and climate change, it also questions whether continued growth can be the definitive goal in a finite sys-

tem—in this case, the island of Catan. By offering ways to win based on responsible environmental stewardship instead of just growth, and by making too much growth potentially lead to all players losing, this board game can

_eo Murray

Perpetually growing hamster after having consumed most of Earth

help players wrestle with the limits of growth.[54]

In the end, whether societal leaders accept it or not, the natural limits of Earth—brought into view by increasing numbers of a population of 7 billion striving to live as consumers—will shatter the myth of continued growth, most likely due to dramatic changes to the planet's systems. Thus degrowth is part of humanity's future. Will people pursue this agenda proactively? Or will Earth and its limits drive the contraction of the global economy?

CHAPTER 3

Planning for Inclusive and Sustainable Urban Development

Eric S. Belsky

Today millions of poor people live in places that are noted on planning maps as "vacant" or "unoccupied." Places that sprang up as unauthorized settlements in developing countries are often still treated by authorities as if they were temporary despite the fact that they are home to nearly a billion people worldwide—people who have invested their labor and meager capital in building their own houses.

These places are by no means temporary, with many families living there for two or more generations, and they are certainly visible. Referred to as slums or informal settlements, many are situated in and around the cores of urban areas. To people who do not live there or draw on the labor, goods, and services supplied by them, these slums are variously viewed as eyesores, illegal settlements, sources of humanitarian concern, home to seemingly intractable problems—at the very least, as places unsafe to visit.

Despite ambivalence on the part of many governments about how and whether to address and plan for these slum areas, international donors, nongovernmental organizations (NGOs), and most governments struggle in a variety of ways to improve conditions in poor communities. Donor agencies and NGOs are especially intent on doing what they can to improve housing and other living conditions by providing services and finance for home improvement and community infrastructure. These agencies and organizations also commonly undertake sectoral efforts to create jobs and build the assets of the poor. More recently, a growing awareness of the threats that climate change poses—especially to slums where environmental risks are acute—has confirmed the importance of creating environmentally sustainable urban regions.[1]

Nonetheless, governments are often wary of putting their limited resources in these communities because land ownership rights in many have yet to be firmly established and because a good deal of the housing and economic activities in slums do not, strictly speaking, conform to laws and regulations. This deepens the ambivalence about how, whether, and which poor communities to invest in. As a result, efforts to address the urban poor just muddle through. Though there are some notable successes, these are set against a backdrop of deep chronic problems that have proved difficult to address effectively, as well as a rapid pace of urbanization that is over-

Eric S. Belsky is managing director of the Joint Center for Housing Studies at Harvard University.

whelming efforts to reduce global levels of urban poverty and poor slum conditions.

Governments must be proactive rather than reactive in addressing slums and the growth of urban poverty. The urban poor are important elements of the urban economy and society. Harnessing their potential to contribute to economic growth and move out of poverty is central to the overall success of national and global economic and social development. Mitigating the environmental and health risks that slum dwellers are exposed to is critical to averting humanitarian crises. Accommodating growth in the numbers of urban poor through deliberate spatial planning—rather than the spontaneous actions of poor migrants—can help avoid new settlements that lack infrastructure and clear and legal title. And addressing the environmental impacts of slums is important to achieving the overall goal of sustainable urban development.

Life in the City

The world's population and economic output are increasingly shifting to urban areas. Over 70 percent of the people in North America, Latin America, and Europe already live in cities. (See Table 3–1.) In Africa and Asia, about 4 out of 10 people are urban. Eastern Africa and South-Central Asia are the least urban subregions of those continents. But Africa and Asia are experiencing the most rapid growth in urbanites. The growth of these populations was 3.4 percent and 2.3 percent a year, respectively, from 2005 to 2010. Both of these regions are

expected to have urban majorities within the next 20 years.[2]

As engines of economic growth, cities are becoming increasingly important to economic output and employment. In fact, the 25 largest cities accounted for roughly 15 percent of the world's gross domestic product (GDP) in 2005. (See Box 3–1.) Levels of development as measured by per capita incomes and GDP appear correlated with the degree of urbanization. In 2005, for example, per capita GDP of predominantly urban western industrial countries was 57 percent higher than in predominantly rural countries. While rapid urbanization was associated with strong income growth in Asia over the past 45 years, however, it did not do so nearly as much in Africa.

Table 3–1. World Urban Population Shares by Major Area and Region, 2000 to 2030

Major Area and Region	Percent of Population in Urban Areas			
	2000	2010	2020	2030
World	46	50	54	59
More developed regions	73	75	78	81
Less developed regions	40	45	50	55
Africa	36	40	45	50
Eastern Africa	21	24	28	33
Middle Africa	37	43	50	56
Northern Africa	48	51	55	61
Southern Africa	54	59	63	68
Western Africa	39	45	51	57
Asia	37	42	47	53
Eastern Asia	40	50	57	64
South-Central Asia	29	32	36	42
Southeastern Asia	38	42	47	53
Western Asia	64	67	69	73
Europe	71	73	75	78
Latin America and the Caribbean	75	80	83	85
Caribbean	61	67	71	75
Central America	69	72	75	78
South America	80	84	87	89
North America	79	82	85	87
Oceania	70	70	70	71

Source: See endnote 2.

Box 3–1. The Rapid Growth of Megacities

The number of megacities—cities with more than 10 million inhabitants—has more than doubled in the last 20 years, from 10 in 1990 to 21 in 2010. They are now home to 7 percent of the world's population. Remaining at the top of the list for the past few years is Tokyo, with 36.7 million people. Delhi, with 22.2 million people, moved from number 11 in 1990 to being the second largest city in 2010. Shanghai's 16.6 million residents moved that city up from being eighteenth in 1990 to seventh in 2010. Forecasts suggest that by 2025 there will be 27 megacities. These high-density cities present a unique set of challenges. But they could also, if planned strategically, offer significant opportunities.

Urban areas use 75 percent of the world's energy, yet dense metropolises provide opportunities for conserving energy and for smarter design. Combined heat and power systems, smart grids, extensive collective transportation, and urban food production can have far-reaching effects.

The provision of vital services, like sanitation and fresh water, is a major challenge in many developing-world megacities. Some 250–500 million cubic meters of drinking water is lost in many of these cities each year. Saving this amount could provide an additional 10–20 million people with drinking water in each city. Delhi has found one solution: managing the supply of water by harvesting rainwater, which has resulted in an actual rise in the groundwater level. Concerns over water supply there still exist, however, as climate change is expected to lead to less rainfall. Moreover, the drinkability of the water in Delhi is an issue that is tightly linked to the provision of sanitation.

Another creative scheme is found in Dhaka, where a partnership among public, private, and civil sectors for dealing with organic waste has yielded many economic, social, and environmental benefits. A composting plant, which can process 700 tons of organic waste per day, was set up with an innovative financing model that made it viable through community involvement and public private cooperation.

Well-designed, dense cities can offer systemic opportunities for mitigation and adaptation to climate change. Mexico City has found that an integrated sustainability plan is effective in tackling air pollution. The areas addressed include land use and planning, transportation, waste management, and climate action planning, as well as seemingly unrelated areas such as water use and supply. Furthermore, the Federal District under the Plan Verde has a regional emphasis on land and ecosystem services conservation, including monitoring and providing clean air, food, forests, and water surrounding the urban area.

Recently there have been new efforts to provide case studies of best practices in sustainable urban planning, such as a United Nations guide for sustainable urban development. Providing more support to urban leaders to implement sustainable development strategies will be essential if megacities are to be part of the solution for a sustainable future.

— Alexandra Hayles
Worldwatch Institute Europe
Source: See endnote 3.

Thus while urbanization may be associated with higher per capita incomes, clearly other factors are also important to broader national income growth.[3]

Urban poverty is pervasive and increasing. According to UN-HABITAT, an estimated 828 million people live in slums, up about 61 million since 2000. With urbanization proceeding in many parts of the world at such a rapid pace, finding ways to provide affordable and decent housing close to jobs is also important to reducing habitat destruction and carbon emissions. The urban poor in slums live in severely impoverished and unhealthy conditions: inadequate sanitation, dilapidated dwellings, poor access to clean water, crowding or insufficient living space, the threat of eviction, inadequate health care, and high exposure to crime. In these areas, health indicators are low and mortality rates are well above average.[4]

Yet as difficult as these conditions are, these communities provide at least rudimentary shelter for large proportions of the urban population. In 2010, slums housed approximately one third of all the urban population in developing countries. At the extreme, an estimated 62 percent of the urban population in sub-Saharan Africa lives in slums. Even in the regions with relatively low shares—such as Latin America—individual countries such as Haiti still have extreme shares of the urban poor living in slums. And because not all the urban poor live in areas classified by the United Nations as slums, the poor's share of the urban population is even higher.[5]

Slums also contain informal sector enterprises such as micro- and small enterprises and home-based ventures that make major contributions to employment creation, local economic development, the urban economy, and national growth. And these activities account for a large share of urban employment. According to the International Labour Organization, for example, 84 percent of all nonagricultural employment in India, 54 percent in Mexico, and 42 percent in Brazil is in the informal sector. Urban consumers and businesses also benefit from the low-cost goods and services produced by the poor. The poor also perform jobs that people in the formal sector would not be interested in but that are necessary for the functioning of the city. And they often play a key role in recycling.[6]

Shifting Attitudes and Priorities

The importance of inclusive and sustainable urban development is coming into clearer view. From scholarly books that draw attention to the importance of the "bottom of the pyramid" and the opportunities it presents for overall development, to specific and reasoned pleas for inclusive and sustainable urban development emerging from the private sector, to the clear growth in interest and institutional arrangements to support sustainable development, attitudes toward urban poverty are starting to change for the better. For instance, a McKinsey Global Institute report on inclusive and sustainable development in India argues that greater attention to the country's growing cities and urban centers is vital to future economic growth and prosperity there. The authors present detailed estimates of the capital expenditure and operating expense needs of the public sector in urban areas across India and recommend how to raise funds to cover them, placing special emphasis on how to manage public land sales and on development regulations to fund safe and decent affordable housing.[7]

The importance of environmentally sensitive or sustainable development is also increasingly being recognized. One indication is that the United Nations, in collaboration with UN-HABITAT, formed the Sustainable Cities Program in the early 1990s. The goal of the program is to design and plan for cities that promote growth and equity by giving priority to environmental sustainability.

Meanwhile, the quickened pace of development and foreign direct investment (FDI) is creating new opportunities to refashion cities while also placing redevelopment pressures on slums that may disrupt their essential shelter and economic functions. According to a 2007 U.N. report, more than 10 million people were forcibly evicted in just six countries from 1995 to 2005. Still, booming economies provide opportunities to tap investment to help pay for slum upgrading. This can be accomplished by harnessing the demand for high-end real estate development to cross-subsidize affordable housing and by converting economic growth and land value appreciation into public revenues for public goods and services. Total FDI in developing nations rose by 12 percent from 2009 to 2010, reaching $574 billion. FDI in real estate and land is also increasing. Real estate development offers high rates of returns in many places because demand for housing and commercial space so outstrips supply. This lends a sense of urgency to addressing slum conditions and the environmental impacts of urban growth.[8]

How Urban Planning Is Currently Conducted

Unfortunately, regional, municipal, and community-level planning seldom takes the poor and their communities meaningfully into account. The most common function of planning at the regional level is to design large-scale infrastructure projects, including water distribution and water disposal, transportation (especially rail and subway), and energy production and distribution. In planning such systems, regional authorities rarely consult community-based organizations in slums. Indeed, with planning maps that often do not acknowledge informal settlements, attending to any disruption caused by infrastructure planning decisions is usually an afterthought. This is especially problematic because the poor often occupy the rights-of-ways where new infrastructure is to be placed.[9]

It is also uncommon for regional authorities to coordinate with municipal authorities and planners until after decisions about the site and types of infrastructure to build have already been made. In addition, regional authorities are explicitly established by national or state and provincial governments to plan and operate large-scale infrastructure. As a result, it is not common for them to consider smaller-scale infrastructure investments that might be critical to or strongly support inclusive and sustainable urban development.[10]

At the municipal level, planning is primarily aimed at smaller-scale infrastructure (such as solid waste disposal or smaller roads), neighborhood development or redevelopment plans, and land use and building regulations. But neighborhood development plans are seldom funded, land use and building regulations often are not widely adhered to, and slum redevelopment plans are produced with little or no community input or public transparency. McKinsey observes, for example, that "on paper India does have urban plans, but they are esoteric rather than practical, rarely followed, and riddled with exemptions." Just as at the regional level, the poor are seldom consulted in a meaningful way in planning for neighborhoods at the municipal level, and many slums are often simply ignored in the process altogether until redevelopment pressures from private interests escalate and motivate municipal authorities to make plans out of the public eye.[11]

Most planning at the municipal level is done in isolation with a narrow sectoral focus—sewer, water, energy, housing, or school facilities, for instance. Yet harmonizing transportation, economic development, land use, and affordable housing goals allows sprawl to be contained, promotes equitable economic growth, reduces traffic and energy consumption, and helps build assets. By contrast, lack of coordination increases the chances of environmental disasters from landslides because poor people often spontaneously occupy slopes. It can also increase carbon dioxide emissions if the poor are pushed to urban fringes far from any work and can cause watershed contamination that affects the city's water supply. In addition, most places fail to make plans for how to house new poor migrants to urban areas. But planning new settlements with decent homes close to jobs that the poor can afford is vital to environmentally sustainable and just cities.[12]

For slum communities, a great deal of planning and investment is aimed at improving conditions and economic outcomes. Though undertaken to some degree by municipal, state, or national governments, international donors and NGOs more often take the lead. Efforts to improve living conditions have focused especially on housing (through efforts to regularize land and to provide subsidies, housing credit, and savings plans to owners so they can finance improvements to their properties) and community infrastructure, such as better roads, solid waste disposal and sanitation systems, potable water distribution systems, health clinics, schools, and spaces for markets, community gatherings, and recreation. An increasing number of slums are demanding that municipal governments provide services such as electricity hookups, water and sewer, and waste disposal.

Sectoral coordination is still not common. However, the likelihood that the views and needs of the poor living in the communities will be taken into account by donors and NGOs in the planning process is much higher than at the municipal level. This is especially so in the case of the actions of community-based organizations. A growing number of these are strong, increasingly active in planning for their communities, and organizing into networks. The most notable of these is Slum Dwellers International. Indeed, a significant amount of donor funding is aimed at helping to form and strengthen community-based organizations and to enumerate the poor.[13]

Strengthening Planning

During the 1950s and 1960s there was great optimism that public sector planning could productively lead and shape national and urban economic and social development. But disillusionment with this model set in during the 1970s—including that planning decisions were nearly always made with little input from the governed.[14]

This ushered in a period of market liberalization and privatization during the 1980s and 1990s. This ideological shift in perspective was promoted by, among others, the U.S. Agency for International Development and the World Bank. The Bank called for governments to create "enabling" frameworks and policies to unleash and channel private investment. As part of the effort to circumvent weak planning capacity at the municipal level, urban development corporations imbued with strong authority, including eviction, were frequently formed. Still, an important feature of the new approach was decentralized government authority and planning.[15]

Like the formal top-down planning approach that preceded it, however, disillusionment with the diminished role of the state set in. Even the World Bank—a key proponent of economic liberalization—acknowledged the importance of public sector planning in a 1997 report, citing studies that such planning in several Asian nations increased economic growth while also achieving a more equitable distribution of its benefits.[16]

A new paradigm for the role of government in managing, directing, and facilitating private investment is starting to emerge. This is a model that once again elevates the importance of public sector planning but approaches it in a way that is both top-down and bottom-up (or participatory) and that tries to facilitate private investment but direct it so that the public purposes of sustainable and inclusive urban development are met.[17]

In short, a strong argument can be made that a reinvigorated public sector must involve all levels of government within a country—national, state/provincial, county, local, and service district. In addition, a truly coordinated effort is required to promote more livable, environmentally sensitive, economically competitive, and inclusive cities. Time and time again, the problems of failing to take a more comprehensive and coordinated view of planning and to engage the poor in the formulation and implementation of plans have limited progress in addressing the needs of the poor and integrating them into economic development.

Combining both strong state planning and a market-enabling framework, governments need to stimulate large- and small-scale private investment, tap into this investment to generate additional public revenues, and channel public and private investments to advance the goals of sustainability, inclusion, and poverty alleviation. While governments should try to leverage and support private investment, they cannot abdicate their role as planners and providers of essential public goods for all communities and income groups. From mega-projects to microfinance and the investments the poor themselves make in housing and microenterprises, private investment shapes the city. It is government's role to encourage and harness this investment to meet important public purposes.

Getting government planners to fully adapt to this new environment will require a great deal of institutional capacity building. It will also take the political will and leadership, often from the national, state, or provincial level, to empower urban planners to coordinate sectoral investments and guide private investment through regulation and public-private partnerships. What would such a "muscular" planning system look like?

First, it would start with the national government formulating a strategy to encourage inclusive and sustainable urban development that is planned and orchestrated by governments and authorities at the urban level. This is because national governments exert a powerful influence over planning and development in urban areas. They control much of the revenue used to fund development at the local level. They grant specific powers to lower levels of government. And they often establish the legal framework for land use regulation, tenure, and ownership rights. Thus it is the national government's responsibility to reform laws and governance structures that discourage rather than encourage inclusive and sustainable development.

While it is still a work in process, one example of an effort to create a national strategy for urban development is the Jawaharal Nehru National Urban Renewal Mission in India. Another broad-based effort that succeeded is Singapore's Urban Development Authority, which addressed overcrowding and urbanization with strategic, multiyear infrastructure and public housing development plans. In 1965, some 70 percent of the city's population lived in overcrowded slum conditions. Today the slums are gone. Narrower national plans that have focused on housing and produced impressive results include Costa Rica's National Housing Finance System and South Africa's 1997 Housing Act and 2010 Breaking New Ground national housing program.[18]

Second, a muscular planning system would feature urban regional planning authorities established by national, state, or provincial governments to coordinate planning across multiple municipalities. Even where there is only a single municipality in an urban area, an authority needs to coordinate across other administrative boundaries like state or provincial, municipal, and service district. In the Mexico City urban region, the Executive Commission for Metropolitan Coordination, established in 1999, addresses regional planning issues in the expanding metropolitan area through a governing body that shares power between the city, Mexico state, and the federal government and that coordinates state and city programs. In China, the national government has orchestrated and integrated regional transportation and economic development in the Pearl River Delta. Targeted spatial policies connect workers and factories to freeways and railways in an overarching polycentric regional plan that leverages urbanization to propel economic development in the region's municipalities.[19]

Third, it would be proactive in planning for growth and change across urban regions. Special attention would be given to the location of new settlements needed to accommodate the poor and to planning for adequate housing and infrastructure. This is necessary to avoid picking locations that isolate the poor from sustainable livelihoods and meeting housing needs in ways that fail to deliver housing and infrastructure the poor can afford. An especially interesting attempt to move from static top-down master plans to a strategic and forward-looking participatory approach was launched in Tanzania in 2000, called the Kahama Strategic Urban Development Planning Framework. It is noteworthy also for its multisectoral nature and for addressing conflicts in development-environment interactions.[20]

Fourth, such a system would produce explicit spatial plans for metropolitan areas. These would plan where new infrastructure would be placed, existing infrastructure improved, affordable housing provided, and business and commercial zones established. It would plan for the physical needs of sectoral interventions such as health clinics and schools for social development, road improvements and public transit for circulation and access, and adequate retail shops, marketplaces, and live-and-work spaces for economic development in slums. These spatial plans would be tied to specific strategies to fund and sequence the needed public investments. Nairobi's Metro 2030 plan captures some of these ideas. In an effort to spur economic development, upgrade sanitation and transportation infra-

structure, and initiate slum upgrades, the city's long-term plan calls for integrated planning approaches that reconcile the multisectoral nature of these challenges.[21]

Modern apartment blocks in Singapore

The urban planning policies in Curitiba, Brazil, are similarly explicitly spatial, and they have led to significant sustainable urban development. The city integrated new Bus Rapid Transit lines with land use plans that stipulated usage and density in order to structure business, commercial, and residential development around the public transportation system. Among other successful aspects of this approach, Curitiba's experience is noteworthy because it reduced traffic congestion, guided urban development, improved air quality, increased citizen mobility, and connected urbanites to housing, employment, and social services across the city.[22]

This need for physical planning extends from large-scale infrastructure planning across urban regions to transportation planning across a city and development plans within residential, commercial, industrial, and mixed-use zones. There are very compelling cases, for example, in which planning for limited redevelopment

Morio

A bus stop shelter under construction in Curitiba, Brazil

and upgrading of slums was far better accommodated by engaging the poor in physical and spatial planning. In several cases, creative solutions to housing and live-and-work spaces that allowed for vertical construction and greater densities were able to free up space for infrastructure improvements and redevelopment while at the same time accommodating displaced residents within the slum in housing and space that met their needs. These include the Ju'er Hutong pilot project in Beijing with its "new courtyard prototype" design; the Walk-Up Kampung Project in Bandung, Indonesia, where families worked with government architects and planners to transform a single-story informal neighborhood into a multistory environment without harming residents' lifestyles by redesigning and reallocating residential and open spaces; and the Favela Bairro squatter settlement upgrading program implemented in Rio de Janeiro in 1993.[23]

Fifth, a muscular municipal planning process would fully engage the poor and community-based organizations in the formulation and implementation of plans and help build the capacity of the poor to participate in planning.

There is a growing consensus that this is critical. In helping the poor create their own vision for the improvement and redevelopment of their communities, it might also be helpful for donors to fund the poor to develop their own plans. One excellent recent example of this is a comprehensive physical plan for the redevelopment of Dharavi in Mumbai, India.[24]

Sixth, it would be fact-based and rooted in detailed information on households, the built environment, municipal service provision, infrastructure, economic activities, environmental conditions and threats, social and community organization, and flows of people and economic activities across parts of the urban region. All this information has yet to be collected in most slums. Increasingly, and productively, the poor themselves are being mobilized to do the collecting. Recent examples of this include the preparation work for the Dharavi plan as well as a project called Map Kibera in Kenya (see Chapter 5) and a digital mapping project of environmental risks in Rio de Janeiro.[25]

Seventh, the planning process would be transparent and accountable. Planning for redevelopment of slum areas is frequently faulted not only for not engaging the poor but for being opaque and having limited accountability for results and the use of invested funds. Indeed, government transparency can enable democratic participation for diverse income groups. In Porto Alegre, Brazil, a 1990s initiative to create a participatory municipal budget engaged diverse community representatives in the city and allowed citizens to scrutinize the municipal budget and allocate its resources according to consensus and need. This degree of participation and transparency enhanced government efficiency, improved waste collection and water delivery for the urban poor, and facilitated the formation of public-private partnerships for the delivery of some municipal services and infrastructure.[26]

Eighth, to the extent possible, planning would be coordinated across sectors but especially in terms of affordable housing, transportation, and economic development. In planning for infrastructure, consideration should be given to whether smaller-scale and distributed infrastructure (such as better roads and bus transit systems rather than large-scale rail system, or solar panels rather than generation and distribution of electricity from plants) might meet certain needs of the poor better than large-scale infrastructure. This occurred, for example, in the development of Bus Rapid Transit systems in Bogota. In addition, many countries may also conclude that an important way to make cities more sustainable will be to try to make them more compact. Indeed, planning for housing, including affordable housing, at high densities along major new transport corridors and in-fill locations is essential for reducing carbon emissions.[27]

Ninth, explicit plans for slums would be developed that take into account their specific situations and that treat the poor fairly. There have been several nationally organized efforts to formulate slum improvement plans, including in Brazil, Colombia, Egypt, Indonesia, Morocco, Mexico, South Africa, Thailand, and Tunisia. In creating such plans, it is important to assess the risks, pressures, and conditions in each slum. A one-size-fits-all policy is impossible. The needs of communities close to employment are different from those further out, for example. It is also important to consider if a slum is at risk from natural disasters like flooding and mudslides. Planners need to consider how much public revenue for slum upgrading could be generated by allowing for partial redevelopment of an area. Although it is difficult to decide to resettle people, such decisions may get made. But in each case provisions must be made for suitable replacement housing, in terms of both minimum or better-quality housing and access to people's livelihoods.[28]

Tenth, with natural disasters an increasing problem in the face of climate change and with urban areas growing so rapidly, planning would also be anticipatory. There are a few good examples of inclusive anticipatory planning to avert the damage that could be caused by natural disasters. These include the Slum Upgrading Facility project in the village of Ketelan in Surakarta, Indonesia, and the Headline Climate Change Adaptation Strategy in Durban, South Africa.[29]

On paper, several efforts around the globe appear to incorporate one or more of these 10 features of a stronger planning and implementation system. But in practice achieving these goals has been problematic, and the results have been inconsistent across developing countries. Mexico, for instance, created the Metropolitan Fund in 2006. Intended to provide federal funding for metropolitan planning challenges and coordination, a 2009 evaluation found that metropolitan areas had consistently allocated their federal funding for roads and highways—which favor private, automobile-based transport—and had scarcely implemented projects for social infrastructure, public transportation, regional economic analysis, or planning for public spaces.[30]

Building this muscular planning system will take time. Its pursuit must not be used as an excuse for inaction on urban poverty and slums. It is vital to pursue every opportunity to improve the living conditions of the urban poor, reduce carbon emissions, and limit habitat destruction created by urban areas and their growth. This means pressing on with sectoral efforts such as placing new affordable housing close to jobs, strengthening community organization and planning, improving public transit to and from slums, conferring security of tenure and land ownership, and building the assets and incomes of the poor. More important, it means not waiting for all preconditions—good institutions, supportive laws and regulations, and well-functioning markets—to be in place but instead to identify, strengthen, and build on efforts that are making some progress so they can be emulated elsewhere.

Barriers to Inclusive and Sustainable Urban Development

Five barriers must be overcome in order to promote inclusive and sustainable urban development. The first is political ambivalence toward improving slums. While on the one hand living conditions in most slums are deplorable and many slum dwellers lack clear rights of ownership, on the other hand the practical reality is that these slums do provide at least rudimentary shelter for large numbers of poor people and are home to businesses that allow them to eke out a subsistence living.[31]

The second major obstacle is the dearth of public resources and private capital to improve the living conditions in slums and help the poor increase their incomes and build assets. The revenue base in most developing countries makes it challenging to allocate enough public resources to make a meaningful dent. The rapid pace of urban growth also presents governments with the difficult choice of directing scarce public resources either to existing slums or to the creation of new authorized settlements that are well planned. This places a premium on finding ways to mobilize private investment in slum improvements, either directly by slum dwellers themselves or by outside lenders, employers, or municipal service providers.[32]

To do so, governments must provide costly public goods: information on economic activity, households, and markets; security of tenure and land ownership; transportation, energy, sanitation, water, and sewer infrastructure; public education and schools; public health facilities; public safety and security; affordable and decent housing for those too poor to secure it on the private market; and insurance to attract private lending and investment when risks are perceived as too great. Even in high-income countries, constraints on public revenues mean these public goods are seldom provided to the full extent needed. But in developing countries, which are at once more pressed for resources and have greater need, these public goods are barely provided at all in slums. This stands in the way of making progress in alleviating poverty.

A third barrier is the conflicts that often arise in making decisions on whether or how to improve, redevelop, or raze slums. Many people benefit from the status quo, directly or indirectly profiting from robust (albeit informal) markets in slums for things like labor, rent, food and other goods, electricity, and credit. These interests may clash with those of people who want to upgrade slums and more deliberately plan and bring public municipal services to them. Conflicts can also arise from ethnic or religious rivalries within slums, class tensions, criminal activity, political corruption, and political and economic pressures to redevelop certain slums. Each of these conflicts is extremely challenging to address. Because the success of slum improvement and planning may depend on their resolution, political leaders may decide they are too intractable. Overcoming this obstacle takes political resolve, conflict resolution skills, and public resources.

The fourth major obstacle is the lack of municipal capacity to make and implement comprehensive plans to alleviate poverty. Even with the political will to address urban poverty, engage the poor in planning, and increase public spending on urban poverty, the governance structures and planning capacities necessary to act effectively are lacking. Municipal agencies are simply unprepared to plan for inclusive and sustainable urban development. Most city governments are scrambling just to make payroll, lack the staff to manage government operations and services well, and do not have any planners—even old-school, top-down master planners, let alone those willing and able to work with multiple actors to advance inclusive planning. When donor agencies attempt to improve municipal planning capacity, there is a strong risk that it will not be sustained when donor funds dry up. Too often there is little if any information on things as basic as how many people live in slums, their demographic characteristics, or where structures, economic activities, social organization, social infrastructure, and public traffic routes are in slums. This makes any serious attempt at spatial planning difficult.[33]

A fifth major obstacle is that the planning profession itself is not well equipped to plan effectively for existing slums. Planning is traditionally aimed at plans for settlement and development before they occur. It involves constructing infrastructure first. Slums stand this sequence on its head in ways that planning methods are not well developed to address—infrastructure must be introduced after a place has been incrementally settled and usually at densities so high that making room for infrastructure involves disrupting current residents and economic activities. This is especially challenging and requires techniques and sensitivities that planners at the municipal level are not trained in and that the profession worldwide is still working on improving.

The fact that so many slums started as unauthorized developments also stands traditional planning on its head. Traditional planning establishes subdivision, zoning, and building controls that are adhered to unless variances are granted for particular, deliberate, planned reasons. In slums, none of these controls were in place early on. It is more challenging to impose them after the fact. Furthermore, traditional planning assumes the rule of law: clear rights of land ownership and tenancy are established, and citizens have recourse to courts to protect their property from unjustifiable or arbitrary

government expropriation. In the case of most slums, these rights are still being negotiated.

Neither engineers nor planners are focused on what infrastructure might best serve the needs of poor residents in urban areas. Infrastructure planning for entire urban regions needs to consider alternatives to large-scale projects that might minimize adverse environmental impacts and better meet the needs of the urban poor. Nor are planners usually trained or accustomed to working across sectors. This tendency is reinforced by institutional structures that funnel funds to sectors separately. So an institutional, technical, and governance capacity to coordinate across sectors is important. This is true at the regional, municipal, and community levels of planning.

Overcoming Barriers

What can be done to clear such formidable barriers so that inclusive and sustainable urban development can become a reality as soon as possible? The answer is to take several bold steps aimed squarely at the identified barriers, equipped with the vision of a stronger, more effective role for the government and community organizations in planning.

National Urban Sustainable Planning and Development Commissions. This action would marshal political will, create a sense of international accountability, and secure the backing of national governments. Establishing these commissions would help elevate the issue of inclusive and sustainable urban development both domestically and internationally, bring a much needed planning focus, incorporate transparency, and allow learning to be shared among nations. It would be important for regional agencies, national universities, and national policy institutes to support and field experts to help set up and staff the national commissions.

The central aims of each commission would be to establish national urban development policies and goals and to modify laws to pro-duce the best division of responsibilities and authorities among levels of government to plan, fund, and implement this development. Each country commission would develop plans that take into account its own resource con-straints, political system, culture, current con-ditions, and market potential. But all commissions would have a common charge that could either be developed at an interna-tional meeting like Rio+20 or could be devel-oped by individual regions or nations. (See Box 3–2.)

National Incentive Funds. Since most metro regions and individual cities have gen-erally failed to create a long-run vision and plan for inclusive and sustainable urban develop-ment, they may need an incentive to do so. A national fund that would cover the costs of set-ting up the appropriate regional and local governance structures, reforming laws, and building the capacity for integrated planning would do this. It would both improve plan-ning and establish more examples that can be studied so that, over time, efforts to improve local planning can be continuously improved. Funding could be tied to demonstrated out-comes to encourage accountability and trans-parency. There is already a model for this at the international level called the Cities Alliance, which is funded by member cities, national governments, and multilateral institutions, including the United Nations and the World Bank. But this is funded presently at a low level. Raising the minimum $100 million per country that is needed for Incentive Funds

---

Box 3–2. Elements of a Charge to National Urban Sustainable Planning and Development Commissions

- Understand how existing laws and policies encourage or discourage inclusive and sus-tainable urban development.
- Gather and map basic information on slum dwellers, economic activities, infrastructure, circulation patterns within and transit access to and from slums, and susceptibility to environmental hazards.
- Assess the potential to use sales of public land and regulation of private development rights to fund improvements to slum com-munities and compensate displaced residents.
- Propose a policy and timeline for land regu-larization in slums to encourage private investment by existing owners and residents and provide them with security from eviction.
- Evaluate the laws and regulations on takings of land and just compensation.
- Establish a plan for strengthening commu-nity-based organizations in poor communi-

ties and engaging them in broader urban planning.
- Establish clear responsibilities and authori-ties at each level of government and plans on how to build governance and govern-ment planning capacity at each level.
- Review what public goods and services governments could provide to encourage private investment in slum housing, infrastructure, services, and businesses.
- Charge urban regional authorities with reviewing infrastructure needs across metro regions, including slums, and considering appropriate alternatives to large-scale infra-structure projects.
- Report back to regional and international bodies in order to share knowledge and best practices as well as to elevate these national efforts and keep peer pressure on govern-ments to make meaningful progress on the goals of their commission.

---

would likely take co-mingling of grant funds from the core budgets of national governments with municipal matching grants, donor assistance, and investments by domestic pension funds and insurance companies.[34]

Financing Innovation Funds. All the best planning will fail to produce results without funds and financing models to support them. National governments should identify, invest in, and export successful financing models. Governments should consider establishing funds that will seed innovations and be used to scale up promising ones in the financing of housing and infrastructure for the poor along with integrated financing tools. Innovative financing vehicles are needed to support incremental slum upgrading as well as larger-scale urban development projects. The few innovative proposals available deserve serious consideration, such as the Kenya Slum Upgrading Project, which involves setting up government-chartered special purpose entities with the authority to issue bonds using, among other things, Crown lands as collateral. Also of interest are innovative approaches to finance subsidized affordable and well-serviced housing as part of large-scale, mixed-use, market-rate developments.

There is also a special need to help slum dwellers finance improvements of their homes. Short of secured real estate lending that requires clear legal title, which most slum dwellers lack, a form of unsecured finance is needed. Unlike microfinance, which has been used with some success to help microenterprises with short-term credit needs, housing credit demands larger and longer-term loans. Underwriting housing loans is quite different. Financing for small community infrastructure is also quite complex because it depends on multiple users paying fees to the international NGO or local community organization arranging for the infrastructure so it can repay the loan. This means that new strategies of finance distinct from established microfinance forms—call them "meso" finance—must be tested. Innovation in product designs, testing of risk assumptions, potential sources of capital, risk-sharing models, and public-private partnerships in the provision of housing and community infrastructure finance is also sorely needed.

There are, though, promising signs that such needs can be met. Encouraging community infrastructure programs include Manila's municipal water administration. The municipal government uses penalties and the prospect of profits to encourage the city's two water concessions to comply with the goal of providing near-universal water service. Accordingly, the concessions have adopted innovative service delivery techniques to reach the urban poor. They no longer require land title for a metered connection, and users can pay for the connection in installments. In addition, users can choose between several metered connections, depending on their income. By 2001, the water concessions, which were created in the mid-1990s, had installed 238,000 new connections, 54 percent of which were in impoverished neighborhoods. On the housing side, a promising case is a community savings and construction program in Mexico called Patrimonio Hoy. Created by Cemex, a global supplier of concrete products, the program has improved housing for participants while providing the company with an adequate return on its investment.[35]

An International Academic Collaborative on Governance and Planning. Because the governance structures and capacity to conduct inclusive and sustainable urban development are presently weak in most countries and urban areas, the attempt to strengthen governance and planning would benefit greatly from an international effort to study best practices, craft and test possible improved governance structures and planning approaches, convene conferences to share knowledge, and develop training programs and planning tools to beef up planning capacity for slums and rapid urban growth in developing countries. It would also be desirable for an academic collaborative to conduct actual interventions to test improvement strategies that are worked out at the local level with government authorities and community-based agencies. The collaborative could also assemble traveling committees of global experts and thought leaders to support governments and planning offices by providing objective, third-party assessments of institutional and legal barriers to inclusive planning and other diagnostics, as well as strategy-building sessions and other technical assistance services for municipal planners.

Small Steps and Bold Actions

Beyond these initiatives, smaller steps could be taken. Some of the steps just described could be pursued even if the larger action were not. Even if national planning commissions were not created, for example, efforts to fund things like community-based organizations to enumerate slum dwellers and build their local capacity to formulate and implement coordinated community development plans could be pursued. Similarly, even if a global academic collaborative were not established, efforts could still be made to fund development of new training materials or a certificate program on inclusive and sustainable urban development at one or more of the leading universities around the world that others could replicate.

The global community can ill afford to ignore the challenges created by urban development and its impact on the environment or by the magnitude and growth of urban poverty. Clearly much must be done, but with the will and a roadmap, the ability to build a brighter urban future for all and for the environment is within grasp. A growing number of examples point the way to better and more-practical planning for inclusive and sustainable urban growth. And there is a surge in interest in the topic. All that is needed now is bold action.

CHAPTER 4

Moving Toward Sustainable Transport

Michael Replogle and Colin Hughes

Danica May Camacho was born in Manila on October 31, 2011—one of a number of children chosen by the United Nations to symbolize the world's 7 billionth resident. Born in one of the fastest-growing megacities in the world, Danica will spend her youngest years in a landscape dominated by cars, jeepney mini buses, heavy trucks, and motorcycles that make it dangerous for her to breathe the air or cross the streets. Manila ranks among the world's worst cities for traffic congestion, commute times, and harmful airborne fine particulate matter from transport sources. In addition, 371 people were killed in traffic in Manila in 2006 alone, and over half of these deaths were of pedestrians. This means that Danica and her parents have an increased risk of respiratory illness and they will spend less time together in their home and more time in traffic. They will also spend a larger portion of their limited income to take motorized modes for trips that are not viable on foot due to unsafe conditions.[1]

Yet these same transport systems also offer important opportunities. They will give Danica and her family access to jobs, markets, and schools. They also provide her city with a way to improve its quality of urban life and lift people from poverty by making its transport infrastructure and services more economically, socially, and environmentally sustainable. The manner in which Manila and thousands of other cities in the developing world manage their transport systems will determine the sustainability of urban life in coming decades for Danica and any children she might have.

World leaders will help shape that future at the June 2012 global summit on sustainable development in Rio de Janeiro. At the 1992 Rio Earth Summit, 187 governments adopted *Agenda 21*, an international action plan on sustainable development that included language supporting sustainable transport. In the two decades since, considerable progress has been made in demonstrating the viability and potential for sustainable transport strategies to meet the mobility needs of growing economies while reducing costs and harm to the environment. But most of the world's transport investments continue to favor unsustainable transport modes. The requisite institutional capacity and governance structures to plan and successfully operate more-sustainable transport systems have not been

Michael Replogle is the global policy director and founder of the Institute for Transportation and Development Policy (ITDP). Colin Hughes is a policy analyst at the Institute.

SUSTAINABLEPROSPERITY.ORG

53

widely developed. Systems to monitor and report on progress toward sustainable transport goals remain weak.

Without changes in policy to mend the trend of unmanaged motorization (see Table 4–1), the outlook for the transport sector is bleak, especially in developing countries. The International Energy Agency (IEA) forecasts that the current number of cars will increase 250–375 percent by 2050, based on various population and economic growth scenarios, while freight activity will also increase 75–100 percent in the same period. The bulk of this growth in transportation activity will happen in the developing world and will impose significant costs to society there. By 2020, road fatalities are projected to rise by 80 percent in low- and middle-income countries. Transportation contributes as much as 80 percent of the harmful air pollutants that cause 1.3 million premature deaths each year, mostly in developing and middle-income countries. And carbon dioxide emissions from transport, an important contributor to climate change, are

expected to grow 300 percent by 2050—with most of the growth again coming from the developing world. This is about five times higher than the minimum reduction of greenhouse gases (GHGs) that the IEA maintains is needed if the transport sector is to meet the Intergovernmental Panel on Climate Change (IPCC) target for avoiding catastrophic climate change.[2]

In the next 20 years the world will see massive growth in demand for transportation fueled by rapid economic development and urbanization. But the current pattern of addressing increased demand for transport—mainly through the expansion of automobile fleets and road network capacity—is unsustainable from economic, social, and environmental standpoints. As former Bogota Mayor Enrique Peñalosa has pointed out, transportation is unique among the problems of the developing world in that it gets worse as a country grows more prosperous. Generally, building new urban high-speed roads and parking capacity for private cars not only fails to

Table 4–1. Characteristics of Unmanaged Motorization and Sustainable Transport

Unmanaged Motorization	Sustainable Transport
Subsidies for motor fuel, parking, and company or government cars	Subsidies for public transport, cycling, and affordable housing close to public transport
Focus on capacity expansion of roads; neglect of local street and sidewalk maintenance	Modernization of roads with real-time traffic management and operations
Motor vehicle traffic and parking displaces cyclists, pedestrians, public transport, parks	Road space protected for pedestrians, cyclists, public space
Disorganized public transport leaves buses stuck in traffic	Bus rapid transit or rail in high-demand corridors, with performance-based contracting
Unmanaged sprawl and urbanization	Public-transport-oriented development
Weak governance structures for transport and land use policy/planning/management	Stronger governance structures for transport and land use policy, planning, and management
Little attention to equality of access among different social and economic groups	More equitable access for the poor, disabled, young, and old

decongest transport networks, it also contaminates urban air, accelerates climate change, increases reliance on imported fuel, and contributes to obesity, respiratory disease, and a growing number of traffic-related fatalities. And it isolates the urban poor, forcing them to choose between low incomes in informal sector employment close to affordable housing and higher-wage jobs that force them to spend a large share of their income and hours each day commuting. But none of this is inevitable. Investments in more-sustainable transport systems can spur more jobs and support more-equitable long-term economic development while protecting the environment.[3]

The Arc of Sustainable Transport in International Agreements

The sustainability challenges facing individual cities and communities—from economic development to climate change—are challenges that are global in scope. They require a framework of commitment at the international level in order to provide incentives for global participation, support global initiatives, and monitor global progress toward goals. In 1992, *Agenda 21* considered transportation a key program area for both resource management and for "improving the social, economic and environmental quality of human settlements." It even went so far as to specifically call for efficient and cost-effective approaches such as integrated land use and transportation planning, high-occupancy public transport, safe cycleways and footpaths, international information exchange, and a reevaluation of present consumption and production patterns. Although transport was featured prominently, however, and even discussed in some depth, no targets, goals, commitments, or other forms of accountability were incorporated.

The Kyoto Protocol adopted by 191 countries since 1997 established legally binding targets for an average reduction of 5 percent of global greenhouse gases relative to 1990 emissions by 2012. With its focus on using markets to find least-cost GHG reduction strategies, it avoided sectoral strategies and did not specifically mention transportation. The climate finance mechanisms it endorsed—the Global Environmental Facility (GEF) and the Clean Development Mechanism (CDM)—were designed primarily around the energy sector, where relatively accurate GHG accounting requires fewer data and is easier to estimate than in the transportation sector. This led to underfunding of sustainable transport projects. While the transport sector now accounts for 27 percent of energy-related GHGs, these climate change mitigation funds have disbursed less than 10 percent of their funding to it.[4]

Although transport is both directly and indirectly crucial to many of the Millennium Development Goals (MDGs), which focus on ending human poverty and were adopted by 193 countries in 2000, transport was scarcely mentioned among the goals and their indicators. The initial recommendations for transport goals as a part of the UN Millennium Project, written by people unfamiliar with the transport sector, were misguided and heavily focused on governmental spending on new road construction. Experts from the World Bank and nongovernmental organizations (NGOs) lobbied to change the recommendations, but the final result was that the UN Millennium Project simply avoided mention of transport. Although it was a blessing that a misguided approach was avoided, ITDP Executive Director Walter Hook noted that "the lack of inclusion of concrete targets for transport in the MDGs carries with it two risks: 1) that critical transport sector interventions will get left off the development agenda entirely, and 2) that the lack of specific targets will give wide latitude to donor agencies and governments to intervene in the sector without any clear guidance from the MDGs, leading to mis-specified interventions that do little to reduce poverty or even make it worse."[5]

The first commitment period under the Kyoto Protocol expires in 2012. In December 2011, the Durban Platform for Enhanced Action was established to present a new plan of action for crafting an agreement to follow Kyoto by 2015. Establishing such a legally binding agreement that includes targets for the world's biggest emitters of GHGs—including the United States, China, and India—in the near term is an essential goal in order to responsibly address the threat of climate change. Another relevant outcome from the Durban summit was the design and structure of a Green Climate Fund that would set up a new system by which industrial countries will help finance implementation of Nationally Appropriate Mitigation Actions (NAMAs) in developing countries. NAMAs are voluntary agreements to reduce GHGs. A key issue in the negotiations is how to design monitoring and evaluation frameworks that enable new funding for NAMA activities in developing countries.[6]

At the moment, this new, bottom-up approach—whereby nations set their own goals for sustainable transportation, receive financing from industrial countries, and cooperate regionally to build capacity and realize goals—represents the most promising pathway to sustainability.

In regards to the transportation sector, several countries have expressed interest in developing transport-specific NAMAs in 2012. Twenty-eight of the 44 NAMA submissions made as of May 2011 specifically refer to mitigation activities in the transport sector. At the same time, a number of leading transport sector NGOs, acting under the umbrella of the Bridging the Gap coalition and the Partnership for Sustainable Low-Carbon Transportation, are working with countries to help them advance this approach.[7]

These efforts have already been advanced through the recent Environmentally Sustainable Transport Forums for Asia and Latin America. The forums resulted in the Bangkok 2020 Declaration, endorsed by 22 Asian countries, and the Bogota Declaration, endorsed by nine Latin American nations. Together with the Report of the Secretary-General to the U.N. Commission on Sustainable Development entitled *Policy Options and Actions for Expediting Progress in Implementation: Transport*, these provide recent evidence of accelerating interest in joint action in this arena. The regional declarations represent a pathway to advance sustainability agreements in a way that avoids the impasse over reduction targets between industrial and developing worlds. But it remains to be seen if these voluntary actions and agreements can engage countries on the wide scale that Kyoto did and achieve the depth of carbon cuts needed for climate stabilization.[8]

Current State of the World: Unmanaged Motorization

Despite growing understanding of the need for sustainable transport, the motorization of the global transport sector has seen unabated growth since at least the 1970s. Recent trends and forecasts of increased growth of vehicle activity in the near future suggest an urgent need to go beyond the status quo approach of linking transport and sustainable development in only a general sense. More-specific institutional development, funding commitments, and accountability frameworks are needed to put transportation on a sustainable path.

Global transport sector energy use has been growing steadily by about 2–2.5 percent a year since 1970 (see Figure 4–1) and is forecast to grow even more quickly in the future. Although the average fuel economy of vehicle engines has improved over time, increases in average vehicle weight, vehicle kilometers traveled, and vehicle fleet size have all led to continued growth in the transport energy consumed and related social costs. In 1990 there were 500 million cars in the world; today there are nearly 800 million, and the IEA forecasts that by 2050 there will be between 2 billion and 3 billion. This means that for every one car stuck in traffic today there will be three or four in 2050. The additional energy use by the transport sector from such rapid growth in vehicles and vehicle activity would far outstrip any reductions from vehicle fuel efficiency improvements, driving transportation energy use even higher.[9]

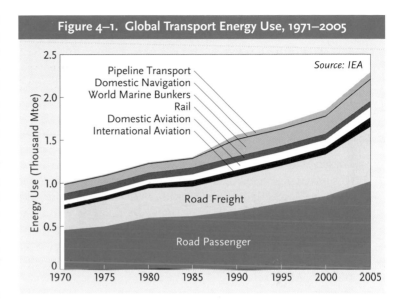

Figure 4–1. Global Transport Energy Use, 1971–2005

Source: IEA

Pipeline Transport
Domestic Navigation
World Marine Bunkers
Rail
Domestic Aviation
International Aviation

Road Freight

Road Passenger

Energy Use (Thousand Mtoe)

(y-axis: 0, 0.5, 1.0, 1.5, 2.0, 2.5)
(x-axis: 1970, 1975, 1980, 1985, 1990, 1995, 2000, 2005)

If current motorization trends continue, the transportation sector will not only help tip the Earth toward catastrophic climate change, it will impose a number of other local economic, social, and environmental costs of motorization. From the health costs related to air pollution to deaths from traffic fatalities and time wasted in traffic, these costs may capture as much as 10 percent of the gross domestic product (GDP) of some countries.[10]

Air Pollution and Public Health. In cities of the developing world, transportation is the source of up to 80 percent of certain harmful air pollutants, including fine particulate matter, carbon monoxide, volatile organic compounds, and lead, as well as nitrous and sulfur oxides. These pollutants can lead to cardiovascular, pulmonary, and respiratory disease as well as various cancers and other illnesses. Threats from transportation-related air pollution are particularly high in developing countries, where less-refined fuels and less-efficient vehicles emit higher levels of pollutants and where a million people die every year from illnesses related to local air pollution.[11]

These health impacts have an economic cost as well. A recent World Bank study on environmental priorities and poverty reduction in Colombia estimated that urban air pollution cost the country $698 million a year due to mortality (65 percent of total cost) and morbidity. The U.S. Federal Highway Administration estimated the total social costs of air pollution associated with U.S. motor vehicle use in 1999 at anywhere from $30 billion to $349 billion a year, mostly associated with premature death and illness caused by particulate matter. While improving air quality requires significant initial investment, the benefits significantly outweigh the costs. A U.S. Environmental Protection Agency study of the Clean Air Act found that between 1970 and 1990 implementation cost $523 billion but the monetized benefits from improved environmental and public health totaled $22.2 trillion. Improved transportation systems combined with air quality regulation could have similar benefits in other countries.[12]

Noise pollution generated by transport can also be detrimental to health and well-being, particularly if it contributes to sleep disturbance, which can lead to increased blood pressure and heart attacks. One study found that the economic cost of noise can reach nearly 0.5 percent of GDP in the European Union.[13]

Congestion. Growth in urban population, income, vehicle fleets, and vehicle travel has in many cities choked road networks. Yet efforts to reduce congestion through expansion of vehicle capacity have been shown to only induce more car travel and increase congestion in the long run. Congestion has many costs: it increases the costs for transport of goods, decreases work productivity, significantly decreases the fuel efficiency of vehicles, increases stress, and decreases the amount of time families can spend together. The Texas Transportation Institute estimates that in 2010, commuters in the 439 U.S. metropolitan areas experienced 4.8 billion vehicle-hours of delay—resulting in 1.9 billion gallons of wasted fuel for a total cost of $101 billion in lost productivity and fuel due to congestion. In the United Kingdom, the estimated cost of time lost in travel is equal to 1.2 percent of GDP. People living in Lima, Peru, are estimated to lose an average of four hours every day in travel, which leads to a loss of approximately $6.2 billion, or around 10 per cent of GDP, every year.[14]

Social Inclusion. Transportation directly affects the places people go and the things they have access to and thus plays an integral role in determining a city's level of equity and social inclusion. The urban poor are particularly vulnerable to the costs of motorized transport while reaping fewer of the benefits because they often cannot afford a car. Without a good public transportation system, the urban poor are further marginalized by their location. This social exclusion affects many aspects of a city-dweller's life, including access to employment, heath care, education, markets, and social and cultural events.

Traditional, auto-focused investments, such as highway and road expansion, tend to benefit the poor the least. Even if public transportation is available, it is often unsafe, expensive, and slow due to congestion caused by private vehicles in mixed traffic lanes. Considerably more public road space is also allocated to car drivers, despite that mode using road space the least efficiently. While a normal bus with a maximum capacity of 50–70 passengers takes up approximately the same amount of space as only three cars with a total average capacity of six passengers, many cities still fail to allocate priority traffic lanes to buses. With 7 billion people and 800 million cars worldwide today, only a minority of people in most of the world have ready access to private motor vehicles. By investing in quality sustainable transportation and giving priority to walking, cycling, and public transport, governments increase social and economic equality and improve the lives of the poor.[15]

Investments that increase car dependence tend to also increase average trip lengths and to put more jobs and opportunities out of reach of the poor. In the United Kingdom, where the length of an average journey has increased by 42 percent since the 1970s, nearly half of the people in the lowest social class report lack of transportation as a barrier to employment. The poorest 20 percent of São Paulo's population spend an average of four hours per day commuting to and from work.[16]

Women also experience social exclusion due to transportation systems. The trips they need to make tend to be off of main public routes, making their transportation more costly in terms of time and money. Additionally, cultural and security factors may restrain women from using certain forms of transportation, such as bicycles, or from riding public transportation after dark.

Road Accidents. The motorization model is also dangerous, especially for the most vulnerable populations. Currently, more than 1.2 million people are killed and 50 million injured every year on the world's roads. Over 90 percent of these deaths occur in developing countries, even though they contain less than half of the world's roads. Today road accidents are the ninth leading cause of death worldwide, but by 2030 they are expected to be the fifth leading cause—greater than deaths from AIDS, lung cancer, diabetes, or violence.[17]

Nearly half of these deaths will be of pedestrians and cyclists killed by drivers. Figure 4–2 illustrates one way that the costs of motorization are disproportionately borne by the poorest segments of society, even though these groups often have little or no access to the mobility benefits from motorization. Vulnerable road users such as cyclists and pedestrians account for 70 percent of traffic deaths in low-income countries, 90 percent of traffic deaths in middle-income countries, and at least 35 percent of deaths even in high-income countries. It is estimated that the global cost of traffic accidents amounts to $518 billion, representing 1–1.5 percent of GDP in low- and middle-income countries and 2 percent of GDP in high-income countries.[18]

In Surabaya, Indonesia, 60 percent of the roads have no usable sidewalks, leading to increased use of motorized transport. For trips of less than 3 kilometers, 60 percent are made by motorized transport. This increases both traffic congestion and the cost to people and businesses that must make more motorized journeys. Investment in sustainable transportation systems and policy changes can make an immediate impact on traffic safety. For example, after implementing the *Transmilenio* Bus Rapid Transit system and *cyclovia* bicycle paths, Bogota, Colombia, saw traffic-related fatalities decrease by 50 percent between 1996 and 2005.[19]

Climate Change. The Intergovernmental Panel on Climate Change's most recent report indicates that in order to limit climate change

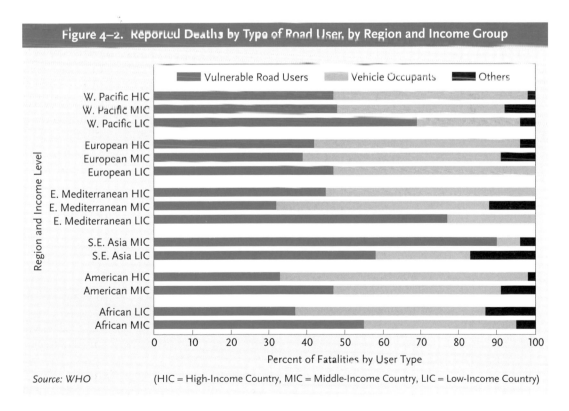

Figure 4–2. Reported Deaths by Type of Road User, by Region and Income Group

Source: WHO   (HIC = High-Income Country, MIC = Middle-Income Country, LIC = Low-Income Country)

to a global average of 2–2.5 degrees Celsius, global GHGs must be cut by 50–85 percent by 2050 (relative to year 2000 emission levels). Several leading climatologists warn that even greater, more immediate GHG cuts may be needed to avoid catastrophic weather events. Given current trends, however, baseline transport GHGs are currently expected to actually increase by 250 percent by 2050. Despite high-level global agreements to promote sustainable transport and reduce greenhouse gases by 5 percent, and despite improvements in sustainable transport technology, planning, and monitoring, the GHGs emitted by transportation have already increased 35 percent since the 1992 Earth Summit in Rio de Janeiro.[20]

Transport is now the fastest-growing source of emissions, and the GHGs associated with all aspects of transportation currently account for 27 percent of global energy-related emissions, as noted earlier. Transport-sector GHG emissions are approaching 10,000 gigatons and growing fast. (See Figure 4–3.)[21]

A recent transport sector assessment by the IEA illustrated how the world could cut transport-sector GHG emissions 40 percent below 2000 levels by 2050 through vehicle and fuel technology and mode-shifting. Several high-level studies suggest that sustainable land use planning, urban design, transportation demand management, and other ways of encouraging low-carbon transport could achieve additional gains while producing net positive user cost savings for travelers. Transportation must undergo major shifts to shape the rate and pattern of motorization, the level of activity of motor vehicle use, and the character of vehicle technology and fuels if it is to contribute reasonably to achieving IPCC targets. Tech fixes alone will not solve the problem.[22]

Targeting a Paradigm Shift in Transport

The good news is that the policies, plans, and technologies that make up this new sustainable transport paradigm have already been identified and proved around the world. They are known as "Avoid, Shift, Improve." They focus on simultaneously avoiding unnecessary motorized trips (with smarter planning, pricing, and telecommunications), shifting trips to more sustainable modes (with sound incentives, information, and investments), and improving vehicle efficiency (with cleaner fuels, better-operated networks, and vehicle technology that is better adapted to individual application environments). Examples of this include Bus Rapid Transit, bike-sharing and cycle-path networks, integrated transit and land use planning, parking limits and pricing, smart parking and car sharing, vehicle registration limits, congestion pric-

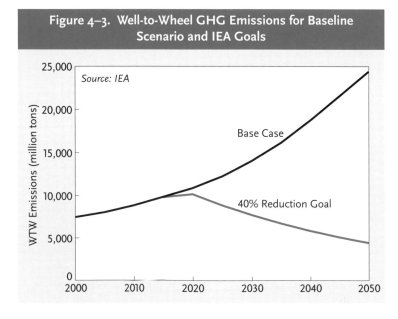

Figure 4–3. Well-to-Wheel GHG Emissions for Baseline Scenario and IEA Goals

Source: IEA

Base Case

40% Reduction Goal

ing, and vehicle emission standards. They include freight logistics and road charging systems, rail modernization, and low-energy freight systems like water and rail. Box 4–1 lists some examples of the Avoid, Shift, Improve approach that have been implemented successfully and shown to reduce transportation user costs, lower emissions, and improve transport service.

These practices also bolster the economy by in many cases creating more jobs, decreasing the time and money wasted in traffic, and achieving independence from costly imported fuels. They generally decrease the local pollutants that cause respiratory illness, reduce obesity through higher activity, reduce traffic fatalities, and lower the emissions of greenhouse gases that cause climate change. When well-managed and taken to scale, sustainable transport can easily accommodate the projected mobility demand related to increases in population, employment, and trade, often at a net negative cost compared with current practices that favor costlier auto mobility.

Transportation is not only a question of energy efficiency and economics, it is an integral part of everyday human life and determines the quality of life in cities. At the same time, transport is highly dependent on the planning and design of cities. Urban development and transportation should first be planned and adapted around the physical scale, needs, and desired lifestyles of citizens—not vice versa. To aid in these efforts, the Institute for Transportation and Development Policy has developed eight principles for transport in urban life. (See Box 4–2.)[23]

Despite its high societal return on investment, sustainable transport faces many barriers to wide implementation. In many countries, the financial and institutional frameworks favor rapid motorization due to specific economic interests, outdated approaches to transportation development, and the dispersal of negative costs to society-at-large instead

> ### Box 4–1. Examples of Best Practices in Avoid-Shift-Improve Approach
>
> **Avoid Unnecessary Motorized Trips**
> - Vehicle registration quotas allocated through auction (Singapore)
> - Congestion charging (London, Stockholm, Milan, Oslo, Bergen, Singapore)
> - Emission-based road use charges for heavy goods vehicles (Germany's national road system)
> - Mixed-use, public transport-dependent development (Curitiba, Hong Kong, Stockholm)
>
> **Shift to More-Sustainable Modes**
> - Bus Rapid Transit (Bogota, Guangzhou, Ahmedabad, Eugene in Oregon)
> - Public bike systems (Paris, Hangzhou, Shanghai, Barcelona)
> - Rail-based mass transit (New York, Hong Kong, Berlin, Tokyo)
> - Pedestrianization, greenways, and cycling networks (Copenhagen, Guangzhou)
> - Parking management and pricing (Zurich, Paris, Tokyo, San Francisco)
> - Intermodal freight system management for optimizing rail and water freight (Germany)
>
> **Improve Vehicle Efficiency**
> - Fuel efficiency regulation (Japan, California, European Union)
> - Electric bikes (20 million+ a year produced in China)
> - High-efficiency cars and trucks: hybrids, neighborhood electric vehicles, biogas buses (Stockholm)
> - Time-of-day road charges (keep traffic at optimal speeds 85 percent of the time in Singapore)

of drivers. This includes domestic public finance, fuel subsidies, official development assistance (ODA) to developing countries, private financial flows, and carbon mitigation financing instruments.[24]

Box 4–2. Principles for Transport in Urban Life

- Walk: Develop neighborhoods that promote walking
- Cycle: Make cycle networks and secure cycle parking a priority
- Connect: Create dense networks of streets and paths
- Transit: Support high-quality transit
- Mix: Plan for mixed uses
- Densify: Match density and transit capacity
- Compact: Create dense regions with short commutes
- Shift: Increase mobility by regulating parking and road use

Source: See endnote 23.

In many countries, a major share of public funds for the transport sector is focused on building roads to support increasing levels of motorized traffic. Subsidies for fossil fuels also claim a significant amount of public funding. These subsidies are socially regressive: the IEA estimates that only 8 percent of the $409 billion that the world spent in 2010 to subsidize fossil fuel consumption (about half of which is used for transport) went to the poorest 20 percent of the population. As the Global Subsidy Institute argues, "while fossil-fuel subsidies are often designed for the interests of poorer populations, they typically benefit medium- to high-income households or lead to diversion. Subsidy reform should be complemented with measures to protect poor and vulnerable groups in society." Additionally, global fossil fuel producer subsidies are estimated to total at least $100 billion annually. Phasing out fossil fuel subsidies would reduce global energy demand by 4.1 percent and carbon dioxide emissions by 4.7 percent by 2020.[25]

ODA flows are also frequently directed toward development based on the motorization model, reflecting both the requests of recipient countries as well as the interests of donor organizations. Financing is particularly directed toward high-value construction and engineering, which overvalues vehicle operating cost savings and undervalues cost-effectiveness, socioeconomic development, and environmental impacts. While some development agencies are improving their planning and transparency toward sustainable transport interventions, cost-effective low-carbon transport is still not a primary goal of assistance.

Private-sector financial flows are also directed toward the development of goods, services, and infrastructure that support the motorization model of transport development, such as motor vehicle manufacturing. One reason is the exclusion of environmental and social costs in the pricing of transport services and vehicles in most countries, which distorts market signals. Regulatory measures, such as emission standards for new vehicles, congestion taxes, carbon taxes, and vehicle registration limits, are currently inadequate in scale and scope to provide a strong signal to the contrary.

Climate mitigation financial instruments such as the GEF and the CDM currently underinvest in carbon mitigation in the transport sector. Funding levels are far from proportionate to the sector's mitigation potential and too limited in scale to catalyze projects. Further, their accounting methodologies, which were designed around the energy sector, are difficult to apply to the transport sector. Emissions from transportation account for over a quarter of all GHG emissions and are the fastest-growing source. Yet much less than one tenth of the cumulative climate change mitigation funds available from the GEF, CDM, and Clean Investment Funds currently goes to the transportation sector, despite the fact that such investments tend to also carry huge co-benefits for local populations in terms of cleaner air, faster travel times, less expensive travel, and more equitable mobility.[26]

While carbon finance typically demands proof of "additionality"—that an investment would not have been made without the availability of the carbon funding—transport investments are almost always made because they produce improved access, economic development, safety, and environmental benefits, and it is carbon reduction that is at best a co-benefit of these primary investment drivers. Moreover, many of the largest impacts of transport investment are indirect, secondary, cumulative, and hard to measure with precision. Nevertheless, the Clean Technology Fund has begun investing in the public transport sector

(see Table 4–2), and the GEF has recently begun to increase transportation sector investment and take a more comprehensive approach to sustainable transport.[27]

Multilateral development banks (MDBs) contribute large flows of capital investment to the transport sectors of developing countries. Investment in transport by the five major MDBs—the African Development Bank, Asian Development Bank (ADB), European Bank for Reconstruction and Development, Inter-American Development Bank, and World Bank—has grown significantly in the last two years, reaching nearly $20 billion in 2010, with continued growth expected thereafter. MDB spending is driven considerably by

Table 4–2. Transport Components in the Clean Technology Fund, March 2010

Country	Investment Cost Transport Component	Total CTF Allocation	Transport CTF Allocation	Transport Components	Emission Reductions from Transport Component
	(million dollars)				$(MtCO_{2eq}$ per year)
Egypt	865	300	100	BRT; light rail transit and rail links; clean technology bus	1.5
Morocco	800	150	30	BRT; tramway; light rail	0.54
Mexico	2,400	500	200	Modal shift to low-carbon alternatives (BRT); promotion of low-carbon bus technology; capacity building	2.0
Thailand	1,267	300	70	BRT corridors	1.16
Philippines	350	250	50	BRT Manila–Cebu; institutional development	0.6–0.8
Vietnam	1,150	250	50	Enhancement of urban rail	1.3
Colombia	2,425	150	100	Implementation of integrated public transit systems; scrapping of old buses; low-carbon bus technologies in transit systems	2.8
Total	9,257	1,900	600		9.9–10.1

Source: See endnote 27.

the types of projects being requested by their member developing countries.[28]

Historically, from the 1970s to 2000, MDB transport sector investment went almost exclusively to building roads for freight and motorized passenger transport. Over the last decade a new approach has taken shape, with action plans, strategic initiatives, and policies on sustainable transport being put in place in different MDBs. Of the $64 billion the MDBs invested in the transport sector from 2006 to 2010, a combined total investment of about $6–7 billion was approved specifically for sustainable transport modes (inclusive of all rail, public transport, nonmotorized transport, and demand management investments). It is expected that in the coming years the portion of MDB funding for road construction will decrease while funding for urban transport, railways, traffic management, and safety will increase.[29]

For example, ADB's 2010 Sustainable Transport Initiative Operational Plan sets a target of investing 30 percent of its transport portfolio in urban transport by 2020 and 20 percent in railways, while reducing road investment to about 42 percent of its portfolio. Within its road operations, ADB—like other MDBs—is emphasizing improved operations and maintenance and rural roads rather than new motorway construction. And recently MDBs have hired more urban transport specialists, railways specialists, and the like rather than traditional road engineers. A joint MDB working group is working toward a common methodology for assessing the GHG impacts of projects they fund. There are discussions between MDBs on road safety, aiming to contribute to the Moscow Declaration on Road Safety and the Global Decade of Action in a harmonized way.[30]

These are welcome changes, but for the MDBs to successfully claim a fundamental reorientation of their transport operations toward sustainable, low-carbon transport they will need to commit more resources in order to create a significant shift to sustainable transport. MDBs will also need to put in place clear criteria for what counts as sustainable transport and set targets for the next decade in consultation with key stakeholders. For instance, not all urban transport is necessarily sustainable. Some types of road investments support sustainability, such as maintenance of existing roads, bicyclist and pedestrian safety improvements, and better traffic management and transit operations. MDBs need to monitor and report publicly on their investments and the impacts of them as well as intensify their efforts to build institutional capacity and partnerships with NGOs, U.N. agencies, and other stakeholders involved with sustainable transport.

Committing to Achieve Sustainable Transport

Despite a long-standing consensus on and understanding of the need for sustainable transport, the lack of clear, transport-specific commitments from the most important stakeholders has largely translated into inaction. New commitments by national governments, MDBs, and other stakeholders to adopt specific sustainable transport goals—with progress measured through appropriate indicators—could help shift the global transport sector to an economically, socially, and environmentally sustainable path.

As a part of any international sustainable development agreements, nations should adopt a transport-specific sustainable development goal or other type of global goal with three targets and appropriate indicators to measure progress toward reducing pollution, facilitating economic development, and promoting equitable transportation:

• ensure global transport GHG emissions and transport sector fossil fuel consumption peak by 2020 and then are cut by 2050 by at least 40 percent below 2005 levels, while ensuring that transport contributes to timely attainment of healthful air quality;

- support the Decade of Action for Road Safety (2011–20) and cut traffic-related deaths in half by 2025; and

- ensure universal access to sustainable transport though support for safe, affordable public transport and safe, attractive facilities for walking and bicycling.[31]

The United Nations should enhance its agency coordination around critical sustainable transport tasks to improve effectiveness in global agenda setting, capacity building, data collection, and cooperation between regions and sectors. It should consider the establishment of a U.N. Transport coordination body to improve its capacity to organize transport sector efforts.

Carbon finance funds, including any future Green Climate Fund, should create a transport-specific financing window to facilitate investment in the sector. This would include transport funding targets commensurate with the sector's share of emissions, adapted impact accounting methodologies without overly restrictive data and modeling requirements, and support for local data collection, monitoring, and institutional development.

National governments, MDBs, and climate funds must also continue to ramp up their engagement with the private sector through public-private partnerships. And they can send the appropriate regulatory signals by working to eliminate subsidies for fossil-fueled vehicles and fossil fuels, adopting polluter pays principles. Fostering multistakeholder partnerships and sharing data with NGOs, civil society, and academia is a key way to build a dynamic and successful shift to sustainable transport.

A formerly congested 10-lane street converted into a multimodal corridor in Guangzhou, China

Wu Wenbin, ITDP

Opportunities for Shifting to Sustainability

The Rio+20 Conference on Sustainable Development presents an important opportunity for the world to make the specific commitments needed to shift the transport sector to a sustainable path. Transport-specific goals as a part of any international agreement will set the stage for global action in this sector and will help foster implementation of sustainable transport even at the neighborhood and city level. These goals can and should continue to guide important initiatives like the Nationally Appropriate Mitigation Actions on climate change submitted by developing countries.

What kind of city will Danica Camacho's children be born into? Will they be able to cross the street safely and breathe healthful air? Will they grow up to get jobs that pay a reasonable wage without wasting hours stuck in traffic? These will be determined by the goals set today and the choices governments make about investing in and managing transportation for tomorrow.

Information and Communications Technologies Creating Livable, Equitable, Sustainable Cities

Diana Lind

The city of Singapore is facing a traffic crisis that costs residents hours every day in missed productivity and gallons in wasted fuel; in Lagos, dangerous building sites injure hundreds of people every year; in Lingrajnagar, India, piped water is available only for a few hours each day, but it is hard for residents to know for sure when that will be. These seemingly disparate urban problems have one thing in common: their solutions come from innovative uses of information and communications technologies (ICTs).

To resolve its traffic jams, Singapore is using cell phone data to map traffic and create alternate travel routes to reduce congestion. In Lagos, a developer has created an app to allow anyone with a smart phone to record the GPS coordinates of a building that looks dangerous and report it to the local government. Even text-messaging technology has a role in improving urban life, as a nonprofit organization in India has shown: it sends texts to people to tell them when their water will be turned on, making it easier to collect water from the tap. In an age when cell phone accounts outnumber people in countries from the United States to Brazil, ICTs are not just connecting people. They are serving as a useful tool to make cities more livable, equitable, and sustainable. (See Figure 5–1 for mobile subscription rates around the world.)[1]

According to the World Bank, 90 percent of urbanization is happening in the developing world, yet most developing countries have Internet penetration levels of less than 50 percent. (But see Figure 5–2 on the increase of Internet users in developing countries since 2006.) As a result, there tends to be a digital divide on a global level: many industrial countries such as the United States and the United Kingdom have cities with robust and varied forms of ICT activity, with most people using broadband Internet, whereas a developing country such as India has just 13 million broadband users. At the same time, strong pockets of ICT activity in countries such as Kenya and digital activism in Arab countries show that the distributed, nonhierarchical nature of the Internet and mobile technologies has removed some barriers to digital inclusion.[2]

While cities might have once been seen as undesirable, unhealthy, dangerous places to live, they are increasingly recognized as criti-

---

Diana Lind is editor in chief of Next American City, a nonprofit that promotes socially and environmentally sustainable economic growth in U.S. cities.

cal assets to nations' economies and environmental strategies. Indeed, their density and infrastructure can make the most of Earth's limited natural resources and their burgeoning populations produce much of the gross world product. Even in India, a country with a rich rural history, Prime Minister Manmohan Singh has said, "If Mumbai fails, then India fails."[3]

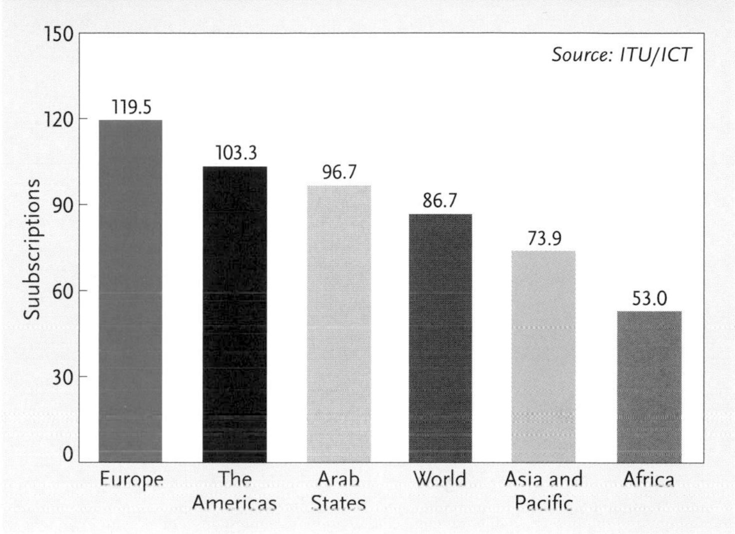

Cities Start to Get Smarter

As cities try to become more sustainable, some municipal governments are finding out just how helpful ICTs can be. Cities run more efficiently when they use an array of intelligent digital infrastructure such as motion-sensor street lamps that save energy and RFID chips in transit passes that allow people to enter a subway or bus with the simple swipe of a card.

Many of these technologies, such as the sensors that enable toll less congestion charging or phone apps that give driving directions with traffic estimates, run on real-time data. As many city governments also collect large amounts of data on residents to help run their health, education, and transportation departments, Web developers are seeking access to this information to create new online solutions for cities. And people around the world are using online tools to participate in virtual social communities that often transform our understanding of real-life ones. These websites and phone apps often encourage civic engagement and dialogue with local government to improve the livability of cities.

Three kinds of actors—local governments, private for-profit or nonprofit businesses, and the public—have largely self-organized accord-

Figure 5–1. Estimated Mobile-Cellular Subscriptions per 100 Inhabitants, 2011

Source: ITU/ICT

	Europe	The Americas	Arab States	World	Asia and Pacific	Africa
Subscriptions	119.5	103.3	96.7	86.7	73.9	53.0

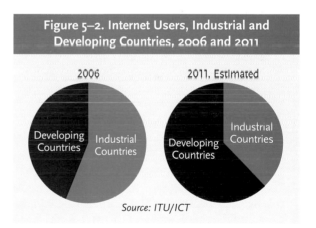

Figure 5–2. Internet Users, Industrial and Developing Countries, 2006 and 2011

2006 — Developing Countries, Industrial Countries

2011, Estimated — Developing Countries, Industrial Countries

Source: ITU/ICT

ing to their strengths and motives in how they use ICTs to improve cities. The private sector has worked with local governments through so-called public-private partnerships to create numerous amenities. Paris's bikeshare system, for example, is made possible by a partnership with the advertising magnate JCDecaux, which provides and maintains the system in return for a substantial share of the city's outdoor advertising space. Other examples include companies paying for transportation infrastructure in exchange for the right to collect tolls on those transportation lines or highways and developers maintaining a public park as part of their private property.[4]

From these standard practices, a new expanded form of public-private partnership has emerged, in which private companies such as Cisco's Smart + Connected Communities, GE's Cities, and Siemen's Sustainable Cities provide technology and products, such as the sensors and equipment required for a remotely controlled energy-efficient smart grid or a conductor-less train system. Additionally, some companies have been contracted as policy analysts, crunching municipal data sets such as crime reports or real estate sales to help governments make decisions on where to allocate police officers or whether to raise property taxes. These varied public-private partnerships have raised a number of questions: Can the chaos and culture of a city be captured in smart algorithms? Should cities outsource some of their responsibilities to corporations?[5]

Many people in the Web development community and urban advocacy movement ask a different question: Do these partnerships and so-called smart cities actually improve the sustainability, livability, and opportunities of urban life? Writing in the *New York Times*, Greg Lindsay, a journalist who has chronicled these developments for numerous publications, notes that "the bias lurking behind every large-scale smart city is a belief that bottom-up complexity can be bottled and put to use for top-down ends—that a central agency, with the right computer program, could one day manage and even dictate the complex needs of an actual city. Instead…the smartest cities are the ones that embrace openness, randomness and serendipity—everything that makes a city great." And Carlo Ratti of MIT and Anthony Townsend of the Institute for the Future observe that "it turns out that sociability, not efficiency, is the true killer app for cities." Peo-

ple rather than technology, serendipity rather than order, may be the saviors of urban life.[6]

Nonprofit organizations and academic programs are working to show that cities can become smarter not by hiring consultants with proprietary tools but by opening their troves of data for public review and use. Nearly every city collects some information on transportation, sanitation, education, wealth, and other indicators. When opened to the public, these data are used to create interactive websites, visualizations, and phone apps that people can use for a variety of purposes, such as finding the nearest bike share station or signing an online petition. These projects have the benefit of making life better for urban citizens and making governance and engagement easier for local officials.

But relying on government data is difficult in places where data collection is inaccurate or the government is uninterested in transparency. In an attempt to address this problem, civic media have emerged that act like social networks for urban advocates. Now people in Los Angeles use a website to alert neighbors of shared service complaints, and people near Fukushima, Japan, who are skeptical of official radiation reports can take their own Geiger counter reports and post the information on a public online map. As interactive as a town hall meeting but often broader in reach, these websites can serve as community watch groups unto themselves.[7]

These actors function best when no one of them is responsible for the direction that ICTs take in cities. But when they collaborate, their top-down, bottom-up, and horizontal distributed approaches have the potential to usher in advances in sustainability, public engagement, and livability in cities.

Beyond Smart Cities

The unprecedented scale and pace of twenty-first century urbanization in Asia offers many countries the opportunity to build new, more-sustainable communities. India alone is expected to need 400 new cities; China will have 221 cities with more than 1 million inhabitants each by 2025. (For comparison, Europe has 35 cities this size today.) Rather than repeat the mistakes of carbon-hogging buildings and car-centric neighborhoods of the past, a handful of cities are serving as experiments in cutting-edge digital and sustainable technology. These "smart" cities are pitched to the public everywhere as more advanced than other cities. But does new technology alone make cities smarter?[8]

Songdo, a 1,500-acre city with 300,000 inhabitants built on landfill in South Korea, exhibits the smart city ethic that combines technologically advanced sustainability measures—a centralized pneumatic waste system that eliminates the need to collect trash and reduces waste by 20–40 percent as well as storm water runoff collection—with low-tech planning principles such as a 100-acre public park and a public transportation system that lets people avoid car ownership. While sustainability is increasingly a priority for Asian cities, Songdo is a noteworthy experiment because of its omnipresent technology. Some 65,000 newly built apartments are equipped with Cisco's home networking systems that will enable people "to conveniently control lighting, air conditioning/heating systems, gas, curtains and all other home devices using touch-screen wall pads, mobile remote controllers and even smartphones, computers and tablet devices," according to the company's website. Cisco TelePresence monitors, which resemble televisions that have video conference capabilities, will allow people not only to participate remotely in a yoga or cooking class but to connect to their local government from the comfort of their homes.[9]

Songdo is far from the only city that is being built from scratch to exhibit digital-sustainable advances. Other examples can be found in India, Russia, China, and the United Arab Emirates. PlanIT in Portugal, a project led by a former Microsoft CEO, is being built by 2015 to accommodate 215,000 people. "PlanIT Valley will combine intelligent buildings with connected vehicles, while providing its citizens with a higher level of information about their built environment than has been possible previously. Its efficiency will extend into the optimum control of peak electricity demand, adapted traffic management for enhanced mobility, assisted parking and providing emergency services with the capacity to have priority when needed in the flow of traffic," the company's website says.[10]

While these cities aim to exhibit drastic improvements in resource management and government efficiency, early visitors report uneasiness with these large-scale innovations. Masdar City, a laboratory for green technologies in the middle of the desert near Abu Dhabi, is testing out the latest in geothermal heating and cooling, electric cars, solar power, and advanced water systems. If successful, it will be the world's first carbon-neutral city. But everything about Masdar, including its eventual population of 50,000, is imported. Nicolai Ouroussoff, the architecture critic at the *New York Times,* noted with disdain that the city resembles a gated community: "[Masdar's] utopian purity, and its isolation from the life of the real city next door, are grounded in the belief—accepted by most people today, it seems—that the only way to create a truly harmonious community, green or otherwise, is to cut it off from the world at large." The United Arab Emirates is one of the world's leading polluters on a per capita basis, and only 1 percent of its energy comes from renewable sources. Is Masdar really a model for the country or just a press-worthy anomaly?[11]

In an attempt to similarly rethink the norm, a small new city called Lavasa was built in India according to New Urbanist principles. (See Box 5–1 for a description of New Urbanism.) But in a country where 830 million people live on less than $2 per day, Lavasa has been chided in the press for its inauthenticity (the name Lavasa was branded by an American company) and the ostentatiousness of reliable water, electricity, and fiber optic connections in each home. Moreover, questions abound about the environmental practices fundamental to creating this "sustainable" city. (The Indian government has sued Lavasa's managing corporation over environmental violations.)[12]

These cities make commendable efforts toward sustainability by using digital technology, but by virtue of their scale, isolation, and private financing, they ignore the basic realities of how most cities function. Rarely do they engage with important stakeholders in cities such as community development organizations, advocacy groups, and education and employment groups that are traditionally involved in the public process. Touted as laboratories, these cities exhibit technologies in an environment that does not easily compare with the "control" of older cities.

Public-private partnerships themselves are not inherently bad for a city. Many projects that provide infrastructure, parks, or other public amenities have greatly enhanced the sustainability of urban life and would not be possible without strong corporate partners. But while they can be very effective in inserting changes to a city, they are best used as part of a city rather than the guiding force behind it.

As an example of how these partnerships can develop, Rotterdam—a city with carbon emissions equal to those of New York, a city 10 times larger—has created a strategic alliance with GE to improve water management, energy efficiency, and emissions reductions in its port area. Using data visualizations, smart meters, and other technologies that form an "intelligent" data system, GE can tell the city how to optimize its energy generation, how to improve the performance of its existing systems, and how to encourage

Box 5–1. Principles of New Urbanism

The Congress for the New Urbanism (CNU) was by cofounded in 1993 by architects Andres Duany, Peter Calthorpe, Elizabeth Moule, Elizabeth Plater-Zyberk, Stefanos Polyzoides, and Dan Solomon to promote walkable, mixed-use neighborhood development, sustainable communities, and healthier living conditions.

CNU is guided by its charter, which supports the principles of:
• livable streets arranged in compact, walkable blocks;
• a range of housing choices to serve people of diverse ages and income levels;
• schools, stores, and other nearby destinations reachable by walking, bicycling, or transit service; and
• an affirming, human-scaled public realm where appropriately designed buildings define and enliven streets and other public spaces.

These communities ideally interweave civic, institutional, and commercial activity in neighborhoods and districts, encourage transit-oriented development, and avoid concentrations of poverty by providing affordable housing.

Some critics complain that these New Urbanist communities can feel inauthentic or overly quaint, given the prevailing contemporary styles of twenty-first-century development. But aesthetics and branding aside, neighborhoods that are walkable, amenity-rich, and less carbon-intensive are crucial to building more-sustainable cities.

Source: See endnote 12.

citizens to monitor their own energy use. GE will also consult on how to make the port area more amenable to residents so as to cut down on the commuting of the 90,000 people who work there. The project will help the city meet its environmental goals of lowering carbon dioxide emissions by 50 percent compared with 1990 levels.[13]

Still, these partnerships are inherently imbued with a top-down ethic that is an increasingly outdated model in today's interconnected, less-hierarchical world. Could there be a way to enhance these projects by including the public in decisionmaking or idea generation? As the Institute for the Future notes in its 10-year forecast, *A Planet of Civic Laboratories: The Future of Cities, Information and Inclusion*: "Industry leaders will have clear visions for the growth of cities—and will promote those agendas with city officials. But the real opportunity for innovation… is inclusive foresight." The key word there is "inclusive." GE's Cities website highlights case studies of how the company is working with cities, proclaiming: "This is our vision for the future and we are working on delivering it today." A city's vision of the future should not be "delivered" by a corporation the same way that it can deliver products. Corporations make fine partners, but not city planners.[14]

The real innovation that local planners need is not just new technology but new ways of engaging the public in the direction and development of cities. Model cities built from scratch with planning, engineering, and technology in mind are likely to lack the character that the public gives to any city and to have unforeseen consequences of their design.

For now, what public-private partnerships lack in terms of the input and ideas of the city's public at large they should give back to the global community in knowledge. Whenever possible, the best practices from these experiences should be shared with the public and other cities. Moreover, there is currently a lack of solid, independent analysis about whether these public-private partnerships are truly achieving their sustainability goals. Independent audits of these projects would vastly contribute to the knowledge about the work being done. Right now these partnership models could be the most—or the least—sustainable way of building cities, but there are no independent data to prove either case.

Last, the financial relationships between cities and corporations are unclear. In 2012, IBM will "give out" $50 million worth of services to cities around the world interested in using technology for management purposes. But what happens after the cities are done with the grant period? It is great to make cities smarter, but at what expense? Given that many municipalities are cutting back on funding for education and employment, cities need to know if technology consultants and public-private partnerships are worth the price.[15]

Data-Driven Cities

More innovative perhaps than the prescriptions offered by consultants and corporations are the new technologies that can engage citizens in understanding better the cities they live in and realizing their own visions for their communities. Replacing the centralized power dynamics of a top-down government or corporate structure, these lateral models take full advantage of the collaborative nature of the Internet and other ICTs.

Data, whether generated by governments or the public, are a central ingredient in empowering and informing people. As a result, the right to data and the use of data have become an increasingly important part of urban policy. Some cities have responded to the growing demand for data by opening their troves of information.

In London, for example, the city opened up 5,400 datasets in the London Datastore. Mayor Boris Johnson said at the Datastore's opening, "I firmly believe that access to information should not just be the preserve of institutions and a limited elite….Data belongs to the people…and getting hold of it should not involve a complex routine of jumping through a series of ever decreasing hoops." One of the largest online warehouses of urban data, the Datastore has opened to public view information ranging from the number of public toilets in London to the languages spoken in schools and the location of vacant commercial spaces. Overlaid on a Google Map or turned into an infographic, these data begin to reveal patterns about the state of the city's infrastructure, education, and economy.[16]

Open data can help urban policymakers shape cities. To give one powerful example, the Spatial Information Design Lab at Columbia University in the United States used rarely accessible data from the criminal justice system to map the home addresses of the country's incarcerated population. The researchers found that a disproportionate number of felons came from a handful of neighborhoods in the largest U.S. cities, meaning that the government was spending more than $1 million a year to incarcerate many of the residents of certain blocks. By revealing this information, the project brought attention to poor housing, education, and health care in these areas. If the condition of these services were improved, could the city reduce its criminal population? Could the government save millions of dollars by investing in the people in these neighborhoods instead of putting them in prisons?[17]

Open data expose aspects of urban life that may not be readily visible, or they help bolster policy initiatives with the facts. Singapore is planning to double its transportation network by 2020. To figure out how to do that, the city has developed a data warehouse that can analyze millions of public transport transactions and recommend routes as a result, helping to make transportation convenient for people. The city has also partnered with the MIT SENSEable City Lab to give people access to real-time information about their city; while this information can be used to see where taxis are during a rainstorm, it can also inform people about energy use changes and how they create a heat-island effect.[18]

Indeed, there are many opportunities to use open data to focus on environmental issues, but surprisingly few of them are being taken. Cities like New York and Washington, DC, have opened hundreds of data sets and held contests for the best apps made from the data. The winning apps in the NYC Big Apps 2.0 contest focused not on environmental sustainability but rather on transit, culture, or governance issues.[19]

If climate change is a priority for these cities, where is the corresponding use of data to inform sustainability policies? Part of the problem may be that environmental data (that is, water and energy usage) are frequently under the control of private companies, but oftentimes open data projects in western countries merely seek to enhance urban life rather than fundamentally change it. Given the environmental, economic, and social crises in many cities, it is time to think about apps that can do more to disrupt the status quo.

The open data movement has a more dramatic impact in cities where the government is beset with bureaucracy. Ushahidi, an open-source technology collaborative, was born out of a need to map violent incidents related to Kenya's elections in 2008. It has since spawned hundreds of websites that visualize information and provide interactive maps of open data. The simply named Budget Tracking Tool, for example, tracks and reports data on how the Kenyan government spends its money; according to the Social Development Network (SOD-NET), officials at the Ministry of Water are under investigation after corruption charges were made by a citizen using the tracking tool. More recently in Kenya—which is 154th out of 182 countries in Transparency International's annual transparency survey—a new open data initiative has made public a relatively modest number of data sets. But symbolically the data have moved the country toward greater government transparency and shifted the way people use the Internet to better their communities.[20]

In India, the Right to Information movement is having a radical effect as the entire country moves toward e-governance. Its Unique Identity program launched in 2010 will eventually identify every Indian by a unique number and biometric data. And while this top-down approach to ICTs will allow many people who formerly had no government identification to gain access to municipal services for the first time and cut down on corruption, it comes with a push to open government data. The Centre for Internet and Society in India is clear that open data have to provide real information and increase the quality of government, not just create fun apps: "It is our belief that open government data in India cannot be as much an issue of providing data for mashing [creating Web apps] and allowing for innovative private-sector information products. Instead, it must be more about addressing the shortcomings of the Right to Information Act…and perhaps moving towards accountability." For cities where private partners are neither available nor desirable, there is an added onus on the government to be transparent before it can act to improve urban life.[21]

Open data and digital transparency's effects are not confined to local governance, however—they will have wide-ranging effects in how the civil sector and even philanthropy work in cities. The International Aid Transparency Initiative, for example, makes information about aid more accessible by requiring donors that are part of the initiative to release information in a common data format. The donors abiding by these standards represent 60 percent of global aid and are thus made more accountable by releasing information on budgets, timelines, activities, and outcomes. The open format allows for millions of dollars in aid to be compared and measured, potentially altering where and how aid is given in slums. Could these same tools be applied to the funding of transportation projects or alternative energy programs to better understand their impact on sustainability?[22]

The plethora of projects created by open data proves that apps are not the only use of municipal information. Nonprofits, media outlets, and others should be as vocal as Web developers in calling for information and finding ways to put data into a broader context with narratives, policy papers, and programs for better urban policy. Moreover, governments should not just open data. They should commit to resolving the problems revealed by the data. Rather than show off the apps built on open data, cities should show how the exposed data affected their policies.

The New Civic Media

In many countries, good sources of open data are hard to get. Governments either cannot or will not release information. At the same time, technology has enabled the public to create more of its own data points and information. With the advent of free, easy blogging software a decade ago, citizen journalists began creating online community forums and blogs. Now, new civic media have emerged that use technology to create a better link between people and government. These media take the form of maps, websites, phone apps, and nonprofit programs and show how the public can use this technology to make their cities more sustainable.

In Kenya, the Kibera slum was left off official maps of Nairobi until 2009. In partnership with an independent team of researchers, Kibera's young people used handheld GPS trackers to create their own interactive map of their area, recording not just streets and buildings, but water pumps, bathrooms, and stores, as well as dangerous or well-lit areas. What resulted was not just a map that could be used by locals but an awareness of Kibera that has spread around the world and caused Kenya to recognize this community of hundreds of thousands. The map in turn launched another media project, Voice of Kibera. Using SMS (short message service), citizens can report where there has been a robbery or fire, as well as start conversations on politics.[23]

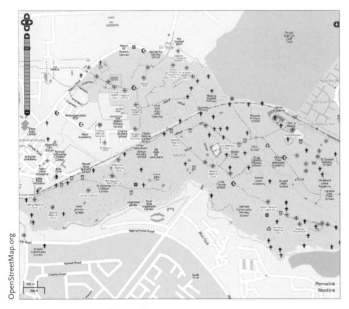

OpenStreetMap.org

Screen capture of the Kibera map

With geocoding, projects like these take community blogging to the next level. In the United States, a veritable grassroots mapping community has emerged. Called The Public Laboratory, a group of activists, educators, technologists, and community organizers develop projects using low-tech materials (balloons filled with helium, simple digital cameras) to take aerial photos and create new maps of contested areas. This kind of work has been helpful in identifying contaminated areas in the Gulf of Mexico near New Orleans after a major oil spill as well as a Brooklyn site of illegal dumping. Like the smart phone app in Lagos that lets people identify dangerous building sites, these projects empower people to direct the government to areas needing attention.[24]

Several websites in the United States and the

United Kingdom are simplifying the interface between citizens and government. FixMy-Street.com in the United Kingdom and SeeClickFix.com in the United States have streamlined and crowd-sourced civic requests and improved on the traditional call centers that many cities use to respond to citizen complaints. SeeClickFix is now directly linked to 311 call centers in many American cities. These projects are remarkable because they expose, if not resolve, the inefficiencies of government's current methods of handling citizens' requests. While advocacy organizations, block captains, and other social groups once organized people, social media are now gathering people online to discuss the future of their neighborhoods and be engaged with city departments. The theory is that once someone knows neighbors are also complaining about an unsafe intersection or the boarded-up house that is used to sell drugs, they will join forces and bring more people to their cause. Indeed, people can vote on which problems are resolved first on SeeClickFix; the better organized a community becomes, the more likely it is that the government will respond.[25]

These media that better connect people and governments can be particularly useful in crisis situations such as earthquakes or war zones. FrontlineSMS, an open-source software "that turns a laptop and mobile phone into a central communications hub," handily proliferates text messages to a group of people. It has been very popular with nongovernmental groups that are providing aid during crises. Similar programs have sprung up to use these technologies for ongoing concerns, not just natural disasters. In India, as mentioned earlier, because water is available for only a few hours a day in many areas, people either waste time waiting for the water or miss it altogether. With drastic improvements in the water infrastructure unlikely in the near future, a nonprofit organization called NextDrop sought to build a reliable mobile network that would notify people by text message when water was on. Initially, NextDrop thought they would pay people small amounts to report the water coming on, spreading the information through their networks. They found out it was better to partner with the utility staff who were turning on the water; in the process, NextDrop became a valuable service to the water utility.[26]

The Hindi word "Jugaad" roughly means a "work-around." A loose translation in the United States might be the scrappy ethos of "do-it-yourself" culture. But rather than doing it by themselves, many tech-savvy organizations are finding ways to "do-it-together" using the inexpensive tools of social media to supplement existing solutions that no longer work. SOD-NET in Kenya launched a new project called Huduma ("service" in Swahili) on the Ushahidi platform that enables people to make requests for better city services via SMS. These reports are then listed online, geocoded, and displayed on a map. Huduma's focus on topics like education, governance, health, infrastructure, justice, and water suggest that this technology can aid cities in solving serious urban concerns. While Huduma's content is still very new, it could present a more comprehensive form of public engagement that moves beyond the 311 model to something resembling a sophisticated online town hall— a bullhorn and watchdog. As a more robust discussion of urban needs, projects like these have greater potential to truly shift policy by creating ambitious targets and open processes.[27]

Civic media have the potential to truly reflect citizens' needs and to enhance government data. Particularly during crises, having technology that allows these voices to be heard is critical. As municipal budgets are squeezed, having citizens empowered to improve their own cities will be critical to their success. These technologies help people help the government but have thus far failed to solve deeper, underlying problems in cities. Instead of fixing a street, how about fixing the education system? Developers of civic media need to strive for these goals of real policy reform, otherwise the technologies risk becoming "nice to have" rather than "need to have."

Turning Information into Knowledge

The types of ICT activity described here have pros and cons. As a result, they work best when in concert with each other. Imagine a city government that works with the business sector to provide technology, which also engages the public through online apps and uses open data to continually assess its progress toward sustainability goals. There is great potential for this kind of arrangement to be successful in improving not just the efficiency of the city but its social and economic character as well.

But as cities increase their use of ICTs for the purposes of efficiency, transparency, and sustainability, investments in technology must be matched by those in the people who lead our cities. Too often, the rhetoric of the smart urban future presents technology as a correction to human inefficiencies. As these examples of ICTs in action show, the best kinds of technology enhance humanity rather than try to eliminate human error. City governments will need to find ways of managing their resources, but they need to do so in a way that recognizes technology as a powerful tool, not a solution in and of itself to the challenges of climate change, economic growth, or social equity.

Take transportation policy, for example. Many cities are struggling to move toward more sustainable transportation systems, which will require huge infrastructure investments. With sensors, cities can implement congestion charging systems, apps can make transit easier to use, data can reveal patterns that will encourage better transportation engineering, and civic media may play a role in determining transportation policy. But to move from car-centered, sprawling cities to multimodal, dense ones, cities need something other than technological change. They need strong leadership. The leadership may come in part from City Hall, but also from local advocates and organizations. Technology will serve these leaders' ideas, and may even shape them, but it cannot replace the vision and power needed to create the big changes that the climate crisis requires.

ICTs such as those described in this chapter can help shift the power dynamic in cities.

In cities where the mayor and other local policymakers have failed to sufficiently reflect citizens' needs, these technologies are helping people fill leadership roles and capture their share of power. They are ensuring that the public's voices are heard and enabling advocates to organize like-minded people. For this reason alone, it is important that governments eliminate as many barriers as possible to Internet access. Accessible technology helps inform and empower people to shape and lead their cities. Using ICTs to build the next generation of organized, informed, and empowered leaders who will use their vision to shape cities— now that's smart.

CHAPTER 6

Measuring U.S. Sustainable Urban Development

Eugenie L. Birch and Amy Lynch

O n Earth Day 2011, Mayor Michael Bloomberg mounted the podium of the Harlem Stage and let a small, triumphant smile cross his face as he announced that New York City was on its way to having the cleanest air of all large cities in the United States. He reminded his audience, who had come to hear the four year progress report on *PlaNYC 2030,* that clean air was a key goal of his much heralded sustainability plan. As of that day, he said, the city had banned Grades 4 and 6 heating oil, a move that, aside from his effort to curb smoking, would be "the single biggest step that we have taken to save lives." Since buildings using low-quality heating oil produce more soot than all the cars and trucks in New York City, cleaning up the air would improve the Big Apple's health record. Every year, he reported, soot caused 3,000 deaths, 2,000 hospital admissions for lung and heart conditions, and approximately 6,000 emergency department visits for asthma in children and adults.[1]

And how could he and other New Yorkers be sure that the city was on track to meet its ambitious clean air goal? They would know by watching two items in *PlaNYC*'s 29-element indicator system that monitored this and the nine other goals in the 2007 plan. First, they would check New York's ranking among U.S. cities (that day it was No. 7—still a long way to go), and then they would take a look at the change in average soot (it was down almost 4 percent from the year before).

This snapshot reflects the work of just one of the more than 200 U.S. cities that have adopted some form of sustainability plan. And in contrast to the majority, which are simply benchmarking or not even measuring their work, it also represents one of the few that have established indicators (or specific metrics) to measure progress. An indicator is a simple measure that signals whether a policy or program is on target to reach a predetermined goal. Analysts distinguish benchmarks—a predetermined milestone to measure progress to a goal—from indicators. For example, Philadelphia's *Greenworks 2009* plan is a benchmarking approach that enumerates aimed-for targets (such as lowering energy consumption in government buildings by 30 percent), not the pursuit of loftier

Eugenie L. Birch is Lawrence C. Nussdorf Professor of Urban Research, Department of City and Regional Planning, and co-director, Penn Institute for Urban Research, University of Pennsylvania. **Amy Lynch** is a Ph.D candidate in City and Regional Planning, University of Pennsylvania.

SUSTAINABLEPROSPERITY.ORG

77

values (such as having the cleanest air of any large city).[2]

While these efforts are all good public policy, the truth is that New York City—like the other U.S. cities—lacks an important marker: it cannot gauge progress against a national standard. The United States does have a sustainable development agenda expressed in the Livability Principles crafted by the Partnership for Sustainable Communities—a federal coalition founded in 2009 when the U.S. Departments of Housing and Urban Development (HUD) and Transportation (DOT) and the U.S. Environmental Protection Agency (EPA) pledged to work together. But it does not have a related indicator system.

This chapter explores the general use of indicator systems in evaluating sustainable development. It focuses on the urban sector of sustainable development for two reasons: first, the United States is 79 percent urban; second, major measures contributing to sustainability necessarily take place at the urban or metropolitan scale.[3]

How to Indicate That Development Is Sustainable

City leaders like Michael Bloomberg are responding to threats of global warming, resource depletion, economic downturns, high levels of poverty, wasteful settlement and urbanization patterns, and a scarcity of adequate, affordable housing and services. They understand that sustainable development is an ongoing process, not a "fixed state of harmony." In their choices of policies and programs, they adhere to the World Commission on Environment and Development's 1987 definition of sustainable development as development that meets the needs of the current generation without compromising the ability of future generations to meet their own needs.[4]

This idea was refined at the 1992 Rio Earth Summit. The key summit document, *Agenda 21*, presented an action plan based on two key values: the removal of disparities (especially poverty) and environmental degradation as well as the integration of environment, social, and economic approaches in order to secure a better future. It explicitly called for monitoring progress through developing indicators: "Methods for assessing interactions between different sectoral environmental, demographic, social and developmental parameters are not sufficiently developed or applied. Indicators of sustainable development need to be developed to provide solid bases for decisionmaking at all levels and to contribute to a self-regulating sustainability of integrated environment and development systems."[5]

In the 20 years since this declaration, much work has been done to strengthen the research, policy, practice, and subsequent evaluation of sustainable development—efforts that will be reviewed at the upcoming United Nations Conference on Sustainable Development (Rio+20) in June 2012. However, some believe that progress has been sluggish, and they attribute the slow adoption of the paradigm to political resistance, limited financial resources, and such technical issues as the absence of scientifically valid and credible indicator systems.[6]

Backed by the call for indicators in evaluating sustainable development, the U.N. Commission on Sustainable Development, which was created to implement *Agenda 21*, has worked for 15 years to develop guidance for interested nations. In a broadly consultative process, it has incorporated evidence-based research from the physical and social sciences to test and refine its recommendations, now in their third iteration, which consist of a suggested list of 50 "core" indicators nested in a larger set of 96. Other entities, including the Organisation for Economic Co-operation and Development (OECD), the Commonwealth Organization of Planners, and a number of Chinese national agencies, are initiating their own indicator systems.[7]

Indicator systems take many forms. Examples of efforts that are focused on sustainable development include the 2005 Environmental Sustainability Index of the Yale Center for Environmental Law and Policy, the European Lifelong Learning Indicators by the Bertelsmann Stiftung, the Sustainable Cities Index by the Forum for the Future in the United Kingdom, the World Health Organization's Indicators to Improve Children's Health, and the European Union's emerging Reference Criteria for Sustainable Cities. Another example is the widely adopted Millennium Development Goals from the United Nations: its eight goals, 18 targets, and 48-element indicator system is representative of the multiple-indicator approach.[8]

In the United States, public and private decisionmakers have long used national indicators or indicator systems to measure important policy goals or progress in particular areas such as life expectancy, gross domestic product, or poverty. Recently, the Office of Management and Budget published 62 social and economic indicators and advocated their use "as quantitative measures of the progress or lack of progress toward some ultimate ends that Government policy is intended to promote" as a means of promoting high-performance government, one whose decisionmaking and policies are based on evidence of "the Nation's greatest needs and challenges and of what strategies are working."[9]

Once established, indicators are often revised and improved. A prime example is the gross domestic product. Since its adoption, international bodies have worked to improve it. Conventions for data collection have been specified in the *System of National Accounts* published by the United Nations, the International Monetary Fund, the World Bank, OECD, and Eurostat, now in its fifth edition after first being published 50 years ago.[10]

Two examples illustrate the range and complexity of indicator efforts. First, the STAR Community Index is an online local-government performance-management tool around issues of sustainability. It is based on 81 goals derived through extensive outreach with local governments and was constructed by ICLEI–Local Governments for Sustainability USA in cooperation with the U.S. Green Building Council, the National League of Cities, and the Center for American Progress. The index encompasses the broad themes of environment, economy, and society, but it does not directly relate to any national policy.[11]

Second, the Green City Index, developed by Siemens in collaboration with the Economist Intelligence Unit, focuses on the environment. It weights quantitative and qualitative measures to rank a limited number of cities worldwide. San Francisco, Vancouver, New York, Seattle, and Denver were named as the most sustainable cities in North America out of a list of 27 considered; in Asia, it was Singapore, Hong Kong, Osaka, Seoul, and Taipei that scored best out of 22 cities.[12]

Establishing indicator systems is most effective when limited to specific purposes. The STAR Community Index has a large number of goals, and it proposes to help individual cities measure performance against each one, not demonstrate progress toward a national standard. The Green City Index focuses only on environmental aspects and highlights only exemplary cities.

While the trend toward awareness of sustainable development is encouraging, the proliferation of indicator systems also presents a number of challenges. With so many systems proposed or in use—each with different goals, objectives, and definitions of sustainable development—understanding broad, national trends is difficult, if not impossible. A better system would be more closely aligned with a stated national agenda.

Despite these limitations, analytical work on indicators of sustainable development has evolved through extensive research and discussion among academics, civic leaders, and practitioners, and it has gained in sophistication, resulting in a robust field of knowledge around the topic. Inquiries have focused on conceptual and definitional issues and on indicator selection, with criteria for assessment. The SMART system, devised by researchers at the Statistical Institute for Asia and the Pacific in Tokyo, tests whether an indicator is specific, measurable, achievable, relevant, and timely. The Institute's researchers also distinguish among pressure, state, and response indicators. Pressure indicators measure actions that may threaten sustainability. State indicators measure current, on-the-ground conditions. Response indicators measure plans and programs that have been undertaken to respond to undesirable states or pressures.[13]

The U.S. Sustainable Development Agenda

Until 2009, the United States lacked a national sustainable development agenda. As a result, many municipalities, some states, several advocacy groups, and a number of private corporations undertook their own sustainable development programs and assessments. But the absence of national guidance meant that their conceptual framing and definitions ranged widely, leaving the United States with no uniform measures for indicating progress toward sustainable development. The basic conundrum is how to align local efforts with a national sustainable development vision and how to establish an efficient, easy-to-apply system for measuring progress toward fulfilling that vision.[14]

In 2009, the federal government decided to address this question when it created the Partnership for Sustainable Communities, which quickly expressed a vision and agenda for sustainable development in six simple statements called Livability Principles. (See Box 6–1.) The principles envision communities at several

---

Box 6–1. Partnership for Sustainable Communities Livability Principles

Provide more transportation choices. Develop safe, reliable, and economical transportation choices to decrease household transportation costs, reduce our nation's dependence on foreign oil, improve air quality, reduce greenhouse gas emissions, and promote public health.

Promote equitable, affordable housing. Expand location- and energy-efficient housing choices for people of all ages, incomes, races, and ethnicities to increase mobility and lower the combined cost of housing and transportation.

Enhance economic competitiveness. Improve economic competitiveness through reliable and timely access to employment centers, educational opportunities, services and other basic needs by workers, as well as expanded business access to markets.

Support existing communities. Target federal funding toward existing communities—through strategies like transit-oriented, mixed-use development and land recycling—to increase community revitalization and the efficiency of public works investments and to safeguard rural landscapes.

Coordinate and leverage federal policies and investment. Align federal policies and funding to remove barriers to collaboration, leverage funding, and increase the accountability and effectiveness of all levels of government to plan for future growth, including making smart energy choices such as locally generated renewable energy.

Value communities and neighborhoods. Enhance the unique characteristics of all communities by investing in healthy, safe, and walkable neighborhoods—rural, urban, or suburban.

Source: See endnote 15.

scales—neighborhood to region—having very different settlement patterns than presently exist in most parts of the United States.[15]

In addition to articulating the Livability Principles, the interagency agreement spelled out a Policy Roadmap to guide the Partnership's future programs. It called for developing a vision and associated supportive definitions for sustainable growth, ensuring integration of the agencies' investment and research activities, and crafting analytical tools to measure progress. (See Box 6–2.) While other U.S. departments and agencies are engaged in sustainable development projects, the Partnership stands out for its clear framing of a specific, comprehensive, and operationalized sustainable development agenda.[16]

The three agencies in the Partnership publicized the Livability Principles in digital and print media, created special offices— HUD's Office of Sustainable Housing and Communities (OSHC) and EPA's Office of Sustainable Communities—to implement programs, began to award funding, built supportive stakeholder groups, and energized regional offices' technical assistance capacities.[17]

In order to advance their Policy Roadmap (especially the goals to enhance integrated planning and investment and to align HUD, DOT, and EPA programs), the Partnership agencies provided funding for a range of programs. At HUD, for example, OSHC created a Sustainable Communities Initiative that issued grants of $100 million in 2010 to 45 localities for regional planning and $40 million in Community Challenge grants to revise local codes to allow coordinated land use and transportation. In 2011, it provided another $97 million for 27 Community Challenge and 29 Regional Planning grants. Due to its loss of program funding in the 2012 budget, OSHC will have to work with other HUD divisions to foster implementation of Livability Principles.[18]

At the Partnership's inception, DOT dedicated $1.5 billion in Transportation Investment Generating Economic Recovery (TIGER) grants to 20 livability projects. In the fall of 2011, when DOT announced a second round of $527 million in these grants for infrastructure investments, it included two important selection criteria: livability (essentially the Livability Principles) and partnership (leveraging other government programs). EPA used a portion of its State Revolving Funds for Water Infrastructure ($3.3 billion) to support livability trials in Maryland, New York, and California and issued Smart Growth Implementation Assistance grants to eight communities that met Livability Principles standards.[19]

The Partnership faces a challenge on its goal to provide a vision for sustainable growth because of the great variety of settlement arrangements—urban, suburban, rural—in the United States. While the U.S. Census identifies the nation's population as being 79 percent urban, 61 percent of Americans live in incorporated places. Of these, more than a quarter live in places with fewer than 25,000 residents, locales where many of the Livability Principles' desired features for housing, land use, and transportation might be difficult to attain. Exemplifying sensitivity to this situation is Transportation Secretary Ray LaHood's endorsement of the Partnership: "Livability means being able to take your kids to school, go to work, see a doctor, drop by the grocery or post office, go out to dinner and a movie, and play with your kids at the park, *all without having to get into your car*. Livability means building communities that help Americans live the lives they want to live—whether those communities are urban centers, small towns, or rural areas" (emphasis added).[20]

Moreover, in the call for transportation alternatives, walkable communities, economic competitiveness, and support for existing communities, the Partnership's sustainable development vision favors urban places characterized by dense, mixed-use settlement patterns under-

Box 6–2. Policy Roadmap for the Partnership for Sustainable Communities

Enhance integrated planning and investment. The partnership will seek to integrate housing, transportation, water infrastructure, and land use planning and investment. HUD, EPA, and DOT propose to make planning grants available to metropolitan areas and create mechanisms to ensure those plans are carried through to localities.

Provide a vision for sustainable growth. This effort will help communities set a vision for sustainable growth and apply federal transportation, water infrastructure, housing, and other investments in an integrated approach that reduces the nation's dependence on foreign oil, reduces greenhouse gas emissions, protects America's air and water, and improves quality of life. Coordinating planning efforts in housing, transportation, air quality, and water—including planning cycles, processes, and geographic coverage—will make more effective use of federal housing and transportation dollars.

Redefine housing affordability and make it transparent. The partnership will develop federal housing affordability measures that include housing and transportation costs and other expenses that are affected by location choices. Although transportation costs now approach or exceed housing costs for many working families, federal definitions of housing affordability do not recognize the strain of soaring transportation costs on homeowners and renters who live in areas isolated from work opportunities and transportation choices. The partnership will redefine affordability to reflect those costs, improve the consideration of the cost of utilities, and provide consumers with enhanced information to help them make housing decisions.

Redevelop underutilized sites. The partnership will work to achieve critical environmental justice goals and other environmental goals by targeting development to locations that already have infrastructure and offer transportation choices. Environmental justice is a particular concern in areas where disinvestment and past industrial use caused pollution and a legacy of contaminated or abandoned sites. This partnership will help return such sites to productive use.

---

pinned by economic agglomeration. Decades of research by urban planners and economists reveal that these elements lend themselves to measurement and evaluation.[21]

So far, the Partnership has underperformed on two important items on the Policy Roadmap: to develop livability measures and tools and to undertake joint research, data collection, and outreach, both of which call for tools to evaluate progress and to establish standardized, efficient performance measures. In the fall of 2011, HUD awarded $2.5 million in research grants in its Sustainable Communities Research Grant Program, but none directly addressed the Policy Roadmap directives with regard to evaluation tools. Conse-

quently, to date the Partnership does not have an easily used set of indicators associated with the Livability Principles. And for cities and regions interested in engaging in federal sustainable development programs, the lack of more generalized standardization is a major drawback, marking an absence of clarity on federal priorities and operations in this area.[22]

Today the Partnership relies on the monitoring systems of individual cities or regions. Yet these may not be comparable and may not include robust measures of sustainable urban development as envisioned in the Livability Principles. The alternative is to rely on less-than-comprehensive federal standards like DOT's and the Office of Management and

Box 6–2. continued

Develop livability measures and tools. The partnership will research, evaluate, and recommend measures that indicate the livability of communities, neighborhoods, and metropolitan areas. These measures could be adopted in subsequent integrated planning efforts to benchmark existing conditions, measure progress toward achieving community visions, and increase accountability. HUD, DOT, and EPA will help communities attain livability goals by *developing and providing analytical tools to evaluate progress* as well as state and local technical assistance programs to remove barriers to coordinated housing, transportation, and environmental protection investments. The partnership will develop incentives to encourage communities to implement, use, and publicize the measures. (emphasis added)

Align HUD, DOT, and EPA programs. HUD, DOT, and EPA will work to assure that their programs maximize the benefits of their combined investments in our communities for livability, affordability, environmental excellence, and the promotion of green jobs of the future. HUD and DOT will work together to identify opportunities to better coordinate their programs and encourage location efficiency in housing and transportation choices. HUD, DOT, and EPA will also share information and review processes to facilitate better informed decisions and coordinate investments.

Undertake joint research, data collection, and outreach. HUD, DOT, and EPA will engage in joint research, data collection, and outreach efforts with stakeholders, to develop information platforms and analytic tools to track housing and transportation options and expenditures, *establish standardized and efficient performance measures*, and identify best practices. (emphasis added)

Source: See endnote 16.

Budget's Scorecard on Sustainability/Energy, an indicator system issued in March 2011 that deals with only part of the sustainable urban development agenda. [23]

National Sustainable Urban Development Indicator Systems

Discussions at a number of international fora have stimulated the Partnership's thinking about how to develop a national system for evaluating sustainable urban development. Following a UN-HABITAT World Urban Forum meeting in March 2010, the White House Office of Urban Affairs and HUD, with support from the Ford Foundation, convened a meeting of stakeholders drawn from the public, private, and nonprofit sectors in the United States and Canada. The intent was to gauge interest in refining North American-oriented approaches to evaluating sustainable development.

The resulting Sustainable Urban Development Working Group agreed on the desirability of an indicator system, but instead of "reinventing the wheel" they opted to rely on, and adapt, indicators that have already proved effective. Representatives from the American Planning Association assembled a list of 22 indicator systems, and the University of Pennsylvania's Institute of Urban Research (Penn IUR) did a literature review and an analysis of existing systems, with the aim of creating a Sustainable Urban Development Indicator Database from which a large portion of the national system could be drawn.

Of the 22 indicator systems reviewed, 8 were created by nongovernmental groups and another 9 by national or municipal governments. Private or professional groups generated

4, and 1 was developed in the academic sector. The International Institute for Sustainable Development has been registering these systems since 1988, and their creation and improvement is accelerating. (See Figure 6–1.)[24]

While some indicator systems apply to more than one scale, 12 of the 22 are city-focused, with 6 also capable of targeting the neighborhood or district level of analysis. Four more account for individual buildings or sites. By comparison, just 2 systems focus at the national level. In terms of substance, 13 systems focus on environmental quality, economic opportunity, and social well-being. But promoting civic awareness, responding to urban migration pressures, and informing municipal investments are also each cited by 6 or 7 systems.[25]

A total of 304 indicators were considered for inclusion in a U.S. Sustainable Urban Development Indicator Database. (This excludes systems that revolve around overly broad goals and objectives or that focus narrowly on highly specific benchmarks.) The challenge was to understand the effectiveness of the remaining indicators in order to match them with the Livability Principles and ultimately select a meaningful and manageable number for a proposed national Sustainable Urban Development Indicator System. The indicators were examined in terms of coverage (environmental, economic, social), "SMARTness," type (pressure, state, response), and breadth (single or multi-dimensional).

Grouping the remaining indicators according to their goals—environmental quality, economic opportunity, and social well-being—shows that all identified systems have several environmental indicators but few social and even fewer economic indicators. Air pollution, environmental stewardship, and water quality or quantity are particularly prominent among the environmental quality indicators. In contrast, no social indicator—such as pub-

Figure 6–1. Sustainable Development Indicator System Timeline

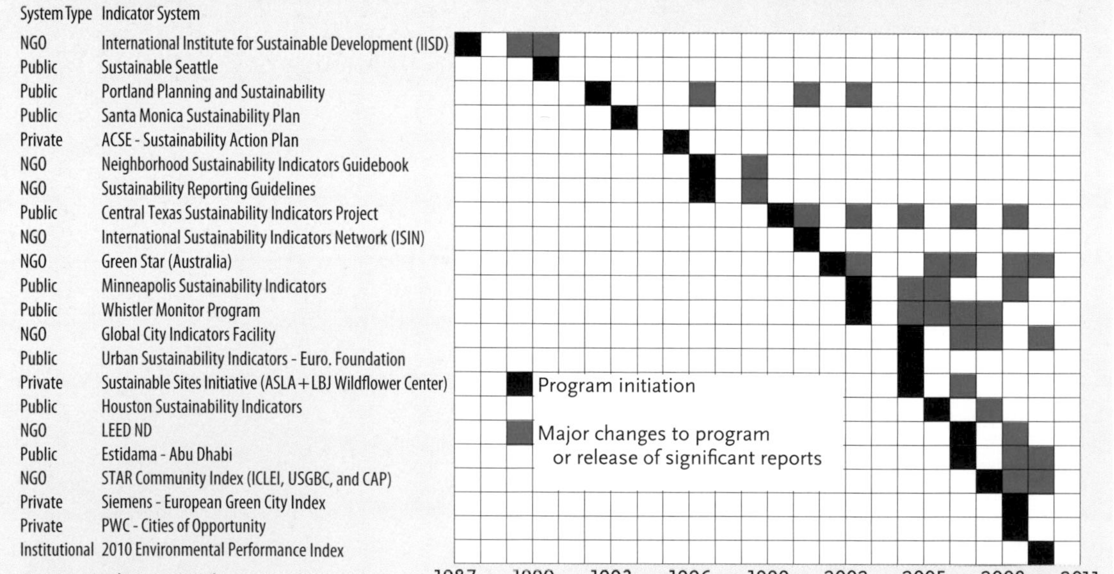

Source: Andreason et al.

lic space, crime levels, or health—appears in more than nine systems, with the majority noted by fewer than five. Only seven economic opportunity goals—such as green buildings, economic competitiveness, and transit infrastructure—are included in the indicator systems. As with the social dimension, none appears in more than nine systems.[26]

Applying the SMART criteria made a dramatic difference, since a large number of indicators were measurable but not achievable, meaning that they asked for information that could be collected but doing so would be prohibitively expensive or difficult. This reduced the number of indicators by more than 50 percent, leaving 145. Of these, 41 percent are social, 34 percent are environmental, and 25 percent are economic.[27]

The type of indictor was also considered, using the pressure, state, or response framework. Paying attention to which types of indicators are included in a system is important because some are more sensitive to different actions than others. For example, a state-oriented system is sensitive to any action that moves the needle in the areas of interest (such as air quality or jobless rate), while a response-oriented system responds only to actions specifically identified in indicators (green buildings constructed, job trainings performed), missing anything not previously specified and overlooking the benefits of the innovative or unexpected. Response-oriented systems may also be of limited duration and must be updated frequently to remain relevant. A suitable indicator system should consist primarily of indicators that describe existing conditions, along with carefully chosen indicators that gauge pressures threatening sustainability and actions taken in response.

The researchers looked for multidimensional indicators as well. Although less common than single-dimensional ones, they are desirable for monitoring important integrative dimensions of sustainable urban development. Fifty percent of the indicators have some degree of multidimensionality, with environmental quality indicators dominating (80 percent), followed by social well-being (37 percent) and economic opportunity (31 percent). If health is considered a facet of social well-being, many existing indicators span environmental quality and social well-being, but far fewer connect environmental quality or social well-being with economic opportunity. This dearth of dual-dimensional indicators and the general lack of economic and social indicators suggests that creating a set of core indicators of sustainable urban development will require going beyond existing systems.

Matching indicators with the six Livability Principles required dissecting their text to determine the themes or types of indicators covered in each principle. (See Table 6–1.) When the researchers linked the indicators in the Sustainable Urban Development Indicator Database to the principles, they found ample evidence that a useful indicator system can likely be created from the database sources, with the exception of the principle to "coordinate and leverage federal policies and investment."[28]

A great deal remains to be done to translate the information from this Indicator Database of 145 indicators into a national Sustainable Urban Development Indicator System. Preliminary results of the next two-year phase in this project will be discussed at Rio+20 in June 2012. The Penn IUR researchers plan to winnow the selection down to 18–20 core indicators through consultation with potential users, pilot testing, and final revisions. A number of key questions will guide this indicator selection process: Is there a valid relationship between the indicator and the Livability Principles? Is the indicator an accurate measure of the items being monitored? Is the indicator sensitive enough to measure progress periodically? What is the correct interval for measuring progress? And is the indicator (and in fact the whole system) cost-effective?

Table 6–1. Livability Principles and Related Indicator Types

Livability Principles	Indicator Types
Provide more transportation choices	Commute mode/mode share Commute time/vehicle miles traveled Carbon emissions
Promote equitable, affordable housing	Housing affordability Equity in housing Housing energy efficiency
Enhance economic competitiveness	Educational attainment Agglomeration Access to credit and capital
Support existing communities	Supporting/revitalizing existing urban areas Promote compact development Conserve and wisely use natural resources Ensure a clean, healthy, and functional natural environment
Coordinate and leverage federal policies and investment	Renewable/locally generated energy State and federal support for local planning efforts
Value communities and neighborhoods	Health Safety Sense of place

Source: See endnote 28.

Returning to another New York City event, when Mayor Bloomberg inaugurated *PlaNYC 2030* to a standing-room only audience in the auditorium of the Museum of Natural History on Earth Day 2007, he did so with firm resolve and a bit of impatience, noting "the science is there. It's time to stop debating it and to start dealing with it." He set the Big Apple on its course. Two years later, with the launch of the Partnership for Sustainable Communities, HUD Secretary Shaun Donovan observed with perhaps a similar tone of impatience: "When it comes to housing, environment and transportation policy, the federal government must speak with one voice." The agendas are there for American cities and the nation. What is missing—but can soon be found, it is hoped—is a national standard to show progress on sustainable urban development.[29]

CHAPTER 7

Reinventing the Corporation

Allen L. White and Monica Baraldi

n early 2011, U.N. Secretary-General Ban Ki-moon challenged the global community with these words: "We need a revolution. Revolutionary thinking. Revolutionary action...It is easy to mouth the words 'sustainable development,' but to make it happen we have to be prepared to make major changes—in our lifestyles, our economic models, our social organization, and our political life."[1]

The Secretary-General is not the first to call for such systemic change. But he, like virtually all others calling for deep change in the planet's development trajectory, left unanswered the critical question of change agent. Who possesses the vision, leadership, and capacity to galvanize the "revolution" leading toward a just and sustainable world during the turbulent decades that lie ahead?

Will leadership come from existing or new global governance bodies equipped with the legitimacy and authority to manage complex and urgent transnational issues such as climate change, responsible international financial regulation, and fair trade? What about civil society? Will it surmount its fragmentation and evolve into a cohesive force for transformational change beyond its contribution to issue-specific

causes such as biodiversity, fair labor practices, and human rights? Is it plausible that a global citizens movement, diffuse and spontaneous yet bound together by common values, forms a social movement that mobilizes millions in support of a "Great Transition"?[2]

And what role for corporations, especially transnational corporations (TNCs) that have a position of global influence that rivals or exceeds the reach of other institutions on the global stage? While corporations by no means stand alone as the sole cause of multiple social and ecological crises, they indisputably play a prominent role in their creation and persistence. Consider the roles of financial institutions in the financial crisis, fossil fuel companies in climate change, and the advertising industry in unsustainable consumerism. Correcting such misalignments will require rethinking the fundamentals regarding societal needs and expectations in relation to corporate form and practices.

Any vision of the global future in the coming decades must include full recognition of the role TNCs play in shaping the planet's human and ecological destiny. It is this reality that animates the intense contemporary debates about the role of business in society and the

Allen L. White is vice president and senior fellow at the Tellus Institute. **Monica Baraldi** of the University of Bologna in Italy is a fellow at the Institute.

capacity—and will—of corporations to simultaneously create public benefit alongside private wealth at a scale and speed commensurate with the needs of a struggling, perilous world. It is difficult, arguably impossible, to imagine a future of 9 billion people living sustainably in the absence of systemic change in the purpose and design of corporations.

Ascendance of Transnationalism

Five centuries ago, the precursor of the modern TNC appeared in the form of sixteenth-century government-chartered trading companies organized by the British and Dutch royalties. While centuries would pass before such global entities would reach the position of dominance now exercised by some 75,000 enterprises, the idea of transnational commerce was set in place by trading companies that functioned as agents of both political and economic domination of the early colonial powers. (See Box 7–1.)[3]

In those early forms, wealth was tied not to the production of goods but to service as brokers between sellers and buyers of spices, silks, minerals, and eventually human beings. The trading companies served to enrich the royalty and, years later, the investors whose capital enabled the expansion of trading activities in return for a share of the profits. The idea that owners of capital could reap the fruits of corporate activity began to take root, the beginning of a slow evolution toward shareholder primacy that centuries later would legitimize "shareholder value" as the core purpose of the modern corporation. Wealth dominated by landholding shifted to wealth accrued through trade enabled by private investors. The era of economic globalization began its slow but steady ascent to full fruition in the post–World War II era.

The march of TNCs toward ever increasing size and geographic reach continues unabated. This trend becomes evident through any num-

ber of measures. Employees in foreign affiliates of TNCs, for example, grew from 21.5 million in 1982 to 81.6 million in 2007. Sales by foreign affiliates increased from $2.7 trillion to $31.2 trillion over the same period—more than an 11-fold increase. Assets increased even more, from $2.2 trillion to nearly $69 trillion.[4]

Figure 7–1 offers a complementary perspective on this expansionist trend. Corporate functions abroad such as sales offices, logistics, call centers, and R&D all show increases between 2008 and 2011. Even corporate headquarters and other "decisionmaking centers" align with this trajectory. Perhaps most ominous from the standpoint of western nations is the movement of R&D functions out of home countries, signaling the growing capacity of emerging economies to participate actively in all aspects of the value chain, not just the traditional resource extraction, processing, and assembly functions long associated with that part of the world.[5]

The first decade of the twenty-first century is thus witnessing dramatic shifts in the global TNC landscape as many corporations seek locations closer to customers and talent in emerging markets. Principal among these is the emergence of TNCs headquartered in these countries, positioning themselves as competitors in the global economy based on scale, technological prowess, and typically a prominent government role in ownership, oversight, and subsidies.[6]

"State capitalism" of this nature is in itself a major force in the expansion of TNCs in Brazil, Russia, and China, driving firms such as SINOPEC, China National Petroleum, State Grid (China), Gazprom (Russia), and Petrobras (Brazil) to rank among the world's top 50 by revenues, with all exceeding $100 billion annually.[7]

As rapid growth in emerging economies continues in the coming decade, the competitiveness of such firms will be a major force in shaping the global future. Along the way, west-

Box 7–1. The Roots of the Modern Corporation

At the dawn of the eighteenth century, industrial production emerged as a new and more powerful source of wealth creation. This development was noteworthy not only for the emergence of the industry-based corporation but also for the shift from wealth dominated by inheritance and social status to wealth driven by entrepreneurship and production of manufactured goods. Innovation gradually displaced entitlement as the primary determinant of such wealth. An entrepreneurial class, fueled by increasing access to private capital, began to redefine the corporate landscape. This process marked the beginning of limited democratization of wealth within the narrow confines of the investor class that a century later would overtake royalty and nobility in terms of aggregate control of the world's wealth.

By the early nineteenth century, two major innovations in corporate form emerged as the dominant architecture that redefined the prevailing model. Until then, partnerships of a few private investors plus the entrepreneur controlled the corporation. In the early 1800s, with opportunities for commerce outstripping the capital available through private partnerships, companies turned to joint stock arrangements. Under this regime, investors large and small could partake in the surging opportunities spawned by industrialization by purchasing stock in the enterprise. Through stock exchanges, investors both near and far could buy shares without substantive involvement—or even knowledge—of how the company operated. Returns in the form of dividends and stock appreciation were enough to attract waves of capital from those seeking to profit from the industrialization that was sweeping Europe and North America.

Along with joint stock ownership, the concept of "limited liability" redefined the nature of the corporation. Limited liability capped risk at a level equal to the value of the investor's shares in the organization, creating the prospect of unlimited gains with limited risk. Industrialists argued before governments that this arrangement was indispensable in order to keep capital flowing to expanding companies that, by the late nineteenth century, were emerging as the world's dominant corporate form.

The dual forces of joint stock and limited liability became the pillars of unprecedented growth in the size, complexity, and profitability of large corporations. The corporation as a remote, tradable asset held by anonymous investors decoupled from management, operations, and community took root. At the same time, labor as a factor of production akin to raw materials whose cost should be minimized became deeply embedded in the world's surging industrial economies.

These attributes put in place the defining characteristic of the modern corporation, namely the primacy of capital (that is, shareholder) interests. The ripple effects of this primacy flowed through every artery of the industrial economy. Conceived as mechanisms to attract money in an era of capital scarcity, shareholder primacy created conditions that would spur many of the social movements that pitted the rights of capital against the rights of labor. With the exceptions of the solidarity during World War II and the shared prosperity of the 1950–80 period, the friction between the rights of capital owners and the rights of labor remain, with varying intensity, a central feature of advanced economies to this day.

Source: See endnote 3.

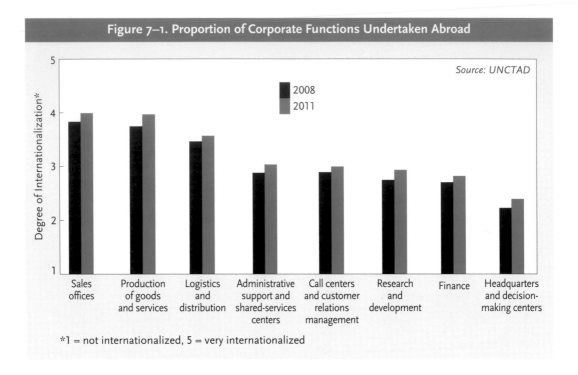

Figure 7–1. Proportion of Corporate Functions Undertaken Abroad

Source: UNCTAD

2008
2011

Degree of Internationalization*

5

4

3

2

1

Sales offices | Production of goods and services | Logistics and distribution | Administrative support and shared-services centers | Call centers and customer relations management | Research and development | Finance | Headquarters and decision-making centers

*1 = not internationalized, 5 = very internationalized

ern standards of corporate governance, social responsibility, and ethics will not automatically become the norms of this new generation of TNCs. Indeed, the world view of the governments and executives that shape such enterprises already is bringing a strong element of sovereign interests in defining what constitutes fair play and responsible conduct in the twenty-first century.

Meanwhile, many other forces render forecasts decidedly uncertain: the availability and price volatility of mineral and food commodities as well as dramatic advances in information technologies and social networking are enabling popular campaigns against whole business sectors and individual companies. The result is rising pressure on business to deliver traditionally public goods such as education and health and, for those firms attuned to new market opportunities, to serve the poorest of the poor with affordable consumer goods and services. Taken together, all signs

point to an era that will challenge sacred cows and prevailing beliefs that underlie the social contract between citizens, their governments, and the corporation—the new party to the bargain.[8]

The Emergence and Limitations of Soft Law

Against this dynamic backdrop, numerous initiatives in the field of corporate sustainability provide a glimpse of the evolving expectations that both reflect and shape new norms of corporate behavior. Most of these fall into the category of voluntary, externally driven efforts to shift corporate behavior toward alignment with the tenets of sustainability: intergenerational responsibility, environmental stewardship, and social justice. This category of initiatives has given rise to a body of "soft law" through which new norms of conduct emerge not through government mandate but through voluntary actions triggered by non-governmental and multilateral actors that, over time, build legitimacy and uptake outside any formal legislative or regulatory process.

Dozens of examples of voluntary initiatives have emerged in the last two decades. As such efforts multiply, many questions have arisen regarding their credibility and impact. Are such programs actually driving corporate conduct toward higher levels of social purpose? Is voluntarism enough to achieve transformational change leading to new norms and measureable outcomes aligned with sustainability? Are these efforts moving away from corporate forms anchored in profit maximization and shareholder enrichment? In short, are such initiatives too incremental and inherently incapable of addressing the urgent and interdependent environmental, social, and economic crises that threaten planetary well-being?[9]

The Global Compact launched in 1999 by U.N. Secretary-General Kofi Annan provides a telling example of the promise and limitations of voluntary initiatives. The Compact operates as a values platform and learning network to advance 10 principles of corporate conduct covering environment, labor standards, human rights, and anti-corruption. It was a historic moment when the world's leading international governmental body explicitly recognized that a just and sustainable future cannot emerge without serious engagement and concrete action on the part of the global business community. Self-described as the "world's largest corporate citizenship and sustainability initiative," some 8,000 participants, including approximately 6,000 corporate endorsers in 135 countries, subscribe to the 10

principles and commit to regular reporting on progress toward their implementation. Among corporate endorsers, about half are large companies and half small or medium-size enterprises (fewer than 250 employees). As measured by company headquarters, the countries most involved in the Compact are France, Spain, and Mexico. (See Figure 7–2.)[10]

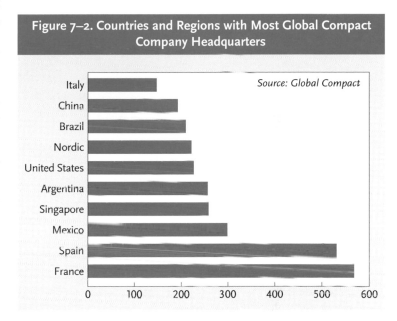

Figure 7–2. Countries and Regions with Most Global Compact Company Headquarters

Source: Global Compact

Notwithstanding its impressive expansion, the Compact has not been immune to criticism for shortcomings in accountability to the U.N. system, inadequate screening and monitoring of participants, and the absence of regular and independent performance monitoring. For example, not until 2004 was a process for "delisting" participants put in place to deal with any failures to prepare the required Communication on Progress (COP). Steady improvement has elevated COP compliance to approximately 76 percent in 2008, a respectable but still flawed level of adherence to the Compact's rules of governance.[11]

A second, kindred effort is the Global Reporting Initiative (GRI). Conceived in 1997 by two U.S. nongovernmental organizations (NGOs)—CERES in partnership with the Tellus Institute— GRI was launched in 2002 at the United Nations as an independent, nonprofit organization affiliated with the U.N. Environment Programme and dedicated to advancement of sustainability reporting by companies worldwide. Approximately 2,000 companies worldwide are registered as users of the GRI Guidelines and countless others do so informally. (See Figure 7-3.)[12]

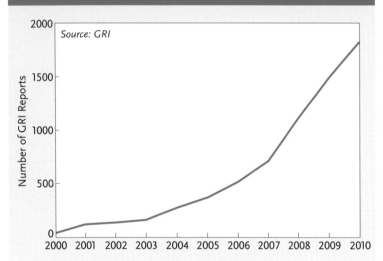

Figure 7–3. Growth in Global Reporting Initiative Reports, 2000–10

As the GRI is adopted more widely, its influence can be seen in dozens of policies, regulations, and programs worldwide. Examples include the Swedish mandate that all state-controlled companies publish GRI-based sustainability reports, the alignment of sustainability performance indicators in the 2011 German Sustainability Code produced by the German Council on Sustainable Development with indicators of the GRI Guidelines, and the listing requirement on the Johannesburg Stock Exchange that corporations comply with the King Code of Corporate Governance, which embraces the GRI framework. Examples of this nature are indicative of a gradual evolution from soft law to semi-mandatory to full-fledged hard law. It is an evolution repeated often in the history of social change, as seen in the suffrage, environmental, women's, and anti-apartheid movements. In all cases, government shifts from observer to participant to codifier of emerging norms whose impulse originates outside of government itself.[13]

Like the Global Compact, GRI is both a reflection and a driver of new norms of corporate behavior. Until its creation, corporate disclosure of environmental, social, and nonfinancial economic impacts was seldom practiced. A handful of firms did so without rules, standardization, or, by and large, credibility. Disclosure initiatives were proliferating throughout the 1990s, advanced by business associations, companies, governments, and NGOs, but they were doing so without an overarching, generally accepted framework built on a core set of principles, processes, and indicators.[14]

In little more than a decade, GRI has been a major force in moving the practice of sustainability reporting from the extraordinary to the exceptional to the expected. In 2013, GRI will release the fourth generation of its Guidelines. Perhaps a half decade hence, sustainability reporting may be blended seamlessly with financial reporting to create a single, "integrated reporting" framework. Meanwhile, the collective experience of thousands of reporters during the last decade is providing the necessary data for investigating the critical but elusive question: Apart from the intrinsic good associated with greater transparency and accountability, is sustainability reporting driving positive change in terms of fair wages, reduced carbon emissions, ethical advertising, and other dimensions of the corporate sustainability agenda? Research during the next few years will shed light on this pivotal question.[15]

Soft law in this field also covers scores of initiatives on specific sectors. SA 8000 on decent work standards overseen by Social Accountability International (SAI), the Marine Stewardship Council, and the Forest Stewardship Council are multistakeholder global efforts seeking to redefine the rules of acceptable corporate conduct, practices, and products. Like the Compact and GRI, these initiatives saw steady expansion during the last decade, measured in terms of certified factories, fisheries, and forest areas.

While progress among voluntary initiatives is undeniable, so too are the limitations. The Compact's 6,000 corporate endorsers, GRI's 2,000 reporters, and SAI's 2,300 facilities are but a tiny fraction of the tens of thousands of TNCs, millions of small to medium size enterprises, and countless factories in the global economy.[16]

Stepping back from individual initatives, the aggregate evidence provides little reason to believe that voluntarism alone is capable of shifting corporate practices at a pace and depth commensurate with the challenges that lie ahead. The bubbles and busts in technology and housing markets throughout many industrial nations, the financial meltdown of 2008 and ensuing recession, intensifying stress and ominous signs of irreparable damage to the biosphere—all these point to an urgent need to move beyond voluntary initiatives and piecemeal, issue-specific, company-driven improvements. Advances in reducing carbon emissions, upgrading supply chain management, and strengthening worker health and safety are noteworthy. But they are no substitute for systemic change that can only occur through deeper reflection centered on the nature of the corporation and the roots of conditions that stand in the way of tapping the corporation's full potential as an agent of sustainable development.

Creating a plausible and inspiring vision of the new corporation requires a set of framing principles and living examples of how such transformative changes would actually look. Such a vision is part and parcel of many current efforts to envision a hopeful planetary future rooted in the values of stewardship, justice, and solidarity on the part of individuals and the institutions that serve them. (See Box 7–2.)[17]

Vision

Imagine the following scenario: In 2015, after years of quiet deliberation and coalition building, an alliance of global business leaders forges an improbable coalition with civil society and labor organizations. Under the relentless pressure of deepening wage and wealth inequalities and intractably high unemployment, and facing an insurgent global citizens movement fueled by a sense of disempowerment and enabled by social networking technology, the alliance steps forward to say:

We are here to declare that business-as-usual is not an adequate response to the expectations, risks, and opportunities for corporations in the twenty-first century. We therefore are advocating a change in the rules governing corporations, a new social contract that recognizes that companies exist at the pleasure of citizens expressed through democratic government processes that provide the rule of law, the stability, and the physical and technological infrastructure upon which all companies depend. The mantra of shareholder value is antithetical to the core values of sustainable development, which is the only long-term pathway to build the prosperous companies and prosperous societies upon which our collective well-being depends.

Box 7–2. Envisioning Sustainable Futures

What will communities and economies look like in the future? Where will current sustainable development policies take the world, if implemented? Most important, what does success look like, if humanity gets things right? The positive vision question is often missing in sustainability discussions, and it is a key reason why shared strategic action is so challenging. Yet as systems thinker Donella Meadows once said, "Vision is the most vital step in the policy process. If we don't know where we want to go, it makes little difference that we make great progress. Yet vision is not only missing almost entirely from policy discussions; it is missing from our whole culture."

When done well, visioning can be very effective, as it both catalyzes creativity and enables people to be far more strategic. The strategic power of visioning lies in how it moves beyond a fragmented, short-term, incremental, and narrow approach. Techniques like scenario planning allow a review of the consequences of the current trajectory and alternative pathways. Having stakeholders think together about the future—an open space still undetermined and thus less burdened by past differences, grievances, and assumptions—makes it easier to reframe stuck debates, build a shared understanding of emerging realities, and identify common interests. Cities like Detroit, countries like

South Africa, and companies like Cisco are all using these collaborative visioning processes to transcend deep divides, create new innovative breakthroughs and strategic flexibility, and discover how to co-design a better shared future for all.

The process of creating a vision draws on the creativity of those involved. Advancing sustainability requires envisioning not only possible futures but positive and compelling visions of these futures. There are an increasing number of initiatives aimed at filling this visioning gap, from the World Business Council on Sustainable Development's Vision 2050 report to the Great Transition Initiative. But supporting a global dialogue—from the grassroots to the mainstream—about the future is still a challenge. The engagement process needs to be deeper, wider, and richer in order to reimagine how humanity can work, live, and play together equitably on Earth.

The good news is that the future is not determined. The gift of foresight means people can anticipate, adapt, and create new alternatives. The negative scenarios can be avoided if people act wisely and boldly now. Indeed, humanity's long-term future as a species may depend on this.

—Nicole-Anne Boyer, Adaptive Edge
Vanessa Timmer, One Earth
Source: See endnote 17.

We commit to creating new global, national, and local governance mechanisms with the authority and resources to encourage and enforce a new generation of corporate accountability and adherence to a new set of principles for corporate design. These principles will provide the beacon for an emergent view of the corporation built on a partnership between people and the biosphere.

A decade ago, this scenario would have been implausible. But in the second decade of the twenty-first century, scrutiny of corporate behavior in both the finance and nonfinance sectors is at an unprecedented level. The volatility and systemic risks during the last decade have shed a harsh light on conduct in the financial sector. Failure to regulate banks' capital requirements properly, the proliferation of exotic and risky derivatives, and the social consequences of the co-mingling of commercial and investment banking activities all conspired to destabilize global financial markets. The widespread public sentiment that the concentration of profits and wealth in the finance sector is a distortionary and unjust force in global markets has corroded confidence in financial organizations and the government entities responsible for their regulation. This has fueled demands for remaking the finance sector to discourage or prohibit financial corporations from becoming "too big to fail" owing to the systemic risk they create in national and global economies. This new situation reflects the fact that the ultimate risk of failure is borne not by investors but by taxpayers of host countries and, in the case of the European Community, a whole region. In these circumstances, "too big to fail" is better characterized as simply "too big."

At the same time, the nonfinancial corporations—the so-called real economy of goods and nonfinancial services—are not immune to diminished public confidence. In industrial countries, a deep and lengthy recession has undermined confidence in the capacity and will of corporations to look beyond the next quarter's share price and earnings to invest in the future of the company, its employees, and the community in which it operates. The gradual "hollowing out" of manufacturing jobs and the threats to traditional safety nets in the United States, much of Europe, and Japan raise profound doubts about corporations' capacity to share fairly the wealth they create among those responsible for its creation. The U.S. government's bailout of the large corporations in the auto industry raised questions about whether such industries should be subject to the same leverage regulations (that is, borrowed debt) and restrictions as are applied to the banking community. And as major emerging economies such as China, Brazil, and India regularly report progress toward poverty alleviation, the contrast with the underperforming, job-deficient industrial nations becomes all the more stark.[18]

At least three conditions are likely to continue to erode public confidence and intensify pressure for change among TNCs: scale, transience, and disparities. First, scale. Unbridled growth continues to expand the market and the political influence of TNCs. Finance, autos, pharmaceuticals, the media, and food are among the sectors that have consolidated into small numbers of major global producers and service providers. The growth imperative, measured by share price, earnings per share, and short-term profits, induces business decisions of questionable societal value: ill-conceived acquisitions to bolster short-term profitability; creative accounting to inflate earnings, for example, through deferral of R&D and maintenance expenditures; executive compensation at historically high levels of disparity relative to median worker wages; and an overreliance on stock options that induces obsessive attention to short-term share price instead of long-term wealth creation. The result is widespread belief that too many TNCs, beholden to shareholder pressure and entrenched short-termism, are failing to contribute to the long term well-being of workers, communities, and the environment at levels commensurate with their capacity to do so.[19]

Transience is a second driver of public disillusionment. In the era of computerized, high-frequency trading, share ownership that years ago was measured in months and years is now reduced to minutes and seconds. Transiency also is manifested in the behavior of footloose industries that continuously seek lower-cost production sites. Accountability to workers and communities has little role to play in this hyper-competitive transient global economy. While the system may yield quick returns to selected investors, the societal costs are severe in terms of diminished community cohesion and worker dislocation. For shareholders, transience is simply the contemporary expression of "creative destruction" that benefits stock traders and sophisticated investors in the name of efficiency and competitiveness. For other stakeholders, the system has diluted the concept of "ownership" to the point where the corporation has become a tradable commodity (like petroleum, minerals, and grains), largely detached from the consequences of its actions on the lives and livelihoods of the communities and individuals where it operates.

Wealth disparities, a third driver of public disquiet with corporate behavior, has never been viewed as the responsibility, much less the outcome, of corporations operating in a free market environment. That role traditionally falls to government. Despite much rhetoric, sustainability issues—climate stability, human rights, poverty alleviation—struggle to attract the attention of both corporate executives and investors. Making the business case for pursuing such issues through investment strategy and portfolio management has made some progress in the last decade, but overall it remains at the margins of managers' and investors' calculus. The results speak for themselves in terms of persistent macro-level income disparities within and across nations, micro-level disparities in the form of the ratio of executive pay to average wages, and record corporate profits juxtaposed with stagnating or declining real wages of workers.

Creating the preconditions of transformational change requires general acceptance of new principles that confront and help reverse the injurious effects of scale, transience, and disparities. Fortunately, multiple initiatives are afoot—focused on both the "new corporation" and the "new economy"—that point to a future in which human and ecological well-being are the centerpieces of a movement toward a just and sustainable future.[20]

Principles

For more than five years Corporation 20/20—an international network with more than 300 participants from business, civil society, law, labor, and academia—has explored the challenges of repurposing and redesigning corporations. Bound by the belief that new models of corporations are indispensable ingredients for a healthy global future, the network developed six Principles of Corporate Redesign as the pillars of its research, advocacy, and public communications. These principles provide an architecture for the next generation of corporations.[21]

- *Principle 1. The purpose of the corporation is to harness private interests to serve the public interest.* Why does society create laws that allow corporations to exist? To serve the public interest, the paramount purpose of all democratic systems. The license to operate is not an entitlement; it is a privilege. It should be granted with terms and conditions aligned with the vision of a just and sustainable society and be subject to periodic review and renewal based on adherence to such vision. This principle recognizes and reinforces the unique capacity of the corporation to generate wealth. At the same time, it insists that in the process of wealth creation, the corporation must act in a manner consistent with the public interest. Where private and public interests conflict, the public interest must prevail. Principle 1 rejects the characterization of the corporation as an insular entity freely marketable without constraints and detached from the broader society in which it operates. Instead, it positions the corporation as inseparable from, and ultimately accountable to, broader societal interests.

- *Principle 2. Corporations shall accrue fair profits for investors, but not at the expense of the legitimate interests of other stakeholders.* Profit and investment are vital to a well-managed corporation. But they are means, not ends. Corporations may not pursue profit for shareholders by undermining the legitimate interests of other stakeholders. The word "legitimate" is critical because claims vary according to the contribution of various stakeholders to the corporate value creation process—specifically, as providers of human, natural, social, and financial capital to the corporation. In the course of corporate activity, off-loading external effects onto society is fundamentally at odds with the public interest. It shall be deemed unacceptable and avoided through appropriate policy and regulatory mechanisms.

- *Principle 3. Corporations shall operate sustainably to meet the needs of the present generation without compromising the ability of future generations to meet their needs.* Vital to the public interest—and vital to human and ecological well-being—is the stewardship of the biosphere through the preservation of natural resources and protection of common assets such as clean air, water, and the Earth's climate. This principle states unequivocally that corporations have intergenerational responsibilities. Managing for short-term private gain is a violation of such responsibilities. Operating sustainably implies a dramatic change in the nature and production of goods and services and the incorporation of the true cost of such production throughout the value chain.

- *Principle 4. Corporations shall distribute their wealth equitably among those who contribute to its creation.* This principle positions equitable wealth distribution as an explicit, though not exclusive, responsibility of business. It rejects prevailing norms of corporate governance and fiduciary duty that make shareholder wealth paramount and the wealth of all other recipients subordinate. Gains to other stakeholders—wages for employees, payments to suppliers, and taxes to local and national governments—currently are defined as costs to be minimized in deference to the primacy of shareholder interests. In contrast, a corporation design aligned with Principle 4 recognizes its obligation to share the wealth it creates equitably among all parties that contribute to it, including payments to gov-

ernment whose provision of stability, security, and the rule of law are indispensable to successful business operations.

- *Principle 5. Corporations shall be governed in a manner that is participatory, transparent, ethical, and accountable.* Participatory governance empowers stakeholders at all levels of corporate decisionmaking. Through governance structures that are transparent and accountable, affected parties are informed, heard, and influential—conditions that nurture productivity, loyalty, and cohesion in the organization. "Stakeholder governance" of this nature is closely intertwined with all other principles. When designed effectively, it is a mechanism for embedding democratic, participatory principles into corporate governance while ensuring that management retains the discretion to operate the organization competitively and efficiently.

- *Principle 6. Corporations shall not infringe on the right of natural persons to govern themselves, nor infringe on other universal human rights.* This principle speaks to the corporation's relationship with the broader political rights of citizens. It delineates a boundary that corporations shall not transgress—namely, the rights of natural persons to govern themselves through delegation of certain rights to government for the benefit of the public interest. Corporations shall not exceed their proper role in democratic political processes and shall respect norms that limit their influence in lawmaking when such influence dilutes or suppresses the voice of the citizenry.

Collectively, these six principles provide the underpinnings of a new corporation whose purpose and design is aligned with the 2015 vision described earlier. The notion that corporations are chartered by governments to serve the public interest harkens back to an earlier era when all corporations—whether royally chartered in the sixteenth century to conduct global trade or state-chartered in the nineteenth to build a road or canal—functioned as time- and scope-limited public purpose entities. It is neither plausible nor necessary to turn the clock back centuries to reconstitute the historical corporate form. But it is both plausible and necessary to rethink such forms in ways that align with the perils and imperatives of inhabiting a sustainable planet in the twenty-first century. That challenge, in turn, requires application of these six principles to key levers of change critical to shaping the corporate "DNA" in the coming decades.

Levers of Change

Corporations are not islands. They are part of a complex economic system with a multitude of simultaneous variables continuously shaping and reshaping their performance. The four dimensions of change described here point to the multiple pathways through which transformational change is possible.

Purpose. Through both legal (charter and bylaws, for instance) and extra-legal (mission statement or family legacy) means, a corporation's statement of purpose both reflects and reinforces its moral fiber. Purpose serves as a touchstone when critical governance, strategy, and policy decisions are on the line. It also serves as a window into the mind of the organization and the degree to which its commitment to creating societal benefit in addition to private wealth is embedded into organizational culture. When the *New York Times* and Novo Nordisk define their purposes, respectively, as "inform the electorate" and "defeat diabetes," it is a signal that social mission plays a significant role in shaping company culture and, ultimately, strategy and management practices.[22]

Countries vary widely in their requirements for organizations to explicitly state their purpose in their bylaws or charter process. In no case, however, is the charter process aggressively deployed as an instrument to advance a country's sustainability agenda. Countries with common law traditions—such as the United Kingdom, the United States, Canada, and Australia—do not legally require a statement of purpose. In contrast, those with civil law traditions typically do, including Germany, France, Italy, Spain, Brazil, and Chile. But even in the latter cases, a declaration of purpose generally means purpose within the context of the business sector in which the firm operates—for instance, "the purpose of company X is to produce pharmaceuticals for institutional and consumer markets."

Recent U.S. developments point to the possibility of a new generation of state charter options that foster a broader, stakeholder-centric statement of purpose. One example is the B Corp (B for benefit), a voluntary alternative to traditional value-free charters that allows companies to use charter language to explicitly recognize the interests of community, workers, the environment, and other non-shareholder stakeholders in operating the organization. B Corps are required to have a corporate purpose to create a material positive impact on society and the environment, to redefine fiduciary duty to require consideration of the interests of employees, community, and the environment when making decisions, and to publicly report annually on their overall social and environmental performance using a comprehensive, credible, independent, and transparent third-party standard. Hawaii, Maryland, Vermont, New Jersey, Virginia, and California have enacted legislation on B Corps as a voluntary alternative to existing shareholder-centric charters, and other states are considering such action. More than 400 companies, mainly start-ups and small organizations, representing over $2 billion in revenue, have been certified as B Corps.[23]

A second charter innovation is California's Flexible Purpose Corporation (FPC), enacted in October 2011. This law aims to provide corporations with a legal framework that protects and encourages a social mission but with a more robust protection of corporate directors from shareholder lawsuits than that contained in B Corp laws. The FPC enables directors to consider the long- and short-term interests of shareholders, employees, suppliers, customers, creditors, community, society, and the environment. Advocates of FPC hope that the new law will attract larger corporations than those drawn to B Corp status by fostering possibilities for mainstream business to blend private and societal wealth creation.[24]

As an instrument for accelerating corporate transformation, charter law is an underused tool. The B Corp and FPC merit the attention of policymakers and regulators in emerging economies at a time when new business formation is accelerating rapidly and the entrenched shareholder-centric fiduciary laws are far weaker than in industrial countries. In emerging economies, charter law could be strengthened by a number of changes aligned with B Corp content: mandatory statement of a public purpose, periodic review of the company's adherence to such purpose as the basis for charter renewal, and mandatory sustainability reporting to disclose specific progress in meeting its self-declared social mission.

Is the spirit of B Corp and FPC transferable to the global stage on which TNCs operate? Yes, it can be and should be. TNCs are global enterprises and, as such, should be subject to enforceable global governance mechanisms much like international trade, international intellectual property rights, and protection of the biosphere (through the Montreal Protocol) already are. A recent proposal for creation of a global chartering entity for TNCs, the World Corporate Charter Organization, offers one approach to achieving congruence between the reach and governance of TNCs. The proposed global charter would complement rather than supplant national charters and would be granted for 10 years. Renewal would be subject to review and confirmation of a TNC's adherence to its charter obligations. A typical global charter would have five parts: a social mission, international norms, an ownership component, a governance section, and an accountability section. Chartering global enterprises at the national or state level represents a fundamental misalignment of purview and regulation that can be rectified only by such a global governance body.[25]

Ownership. Ownership, like purpose, plays a pivotal role in shaping and reinforcing the worldview of the corporation. Many alternatives to the dominant western ownership model of the joint stock, limited liability corporation—such as trust ownership, employee ownership, cooperative ownership, and community-based, hybrid social enterprises—are generally better aligned with the principles of corporate design described earlier.

Such corporate forms readily mesh well with concepts such as stakeholder governance, fair distribution of wealth created by the enterprise, and orientation to long-term horizons. Moreover, such forms are not curiosities. They in fact exist by the thousands in numerous countries, though their prominence is seldom tied to broader debates surrounding business-society relations. In the United States alone there are some 11,000 totally or partially employee-owned companies and 130 million members of urban, agricultural, and credit union cooperatives. In Europe, over 300,000 cooperative businesses employ 5 million people. In Spain, the Mondragon cooperative in the Basque region is a prosperous, umbrella entity with 100,000 employees in a wide range of product and service enterprises. In Italy, the Lombardia region counts over 11,500 cooperatives and 170,000 related employees. And in the United Kingdom, the John Lewis Partnership—a $10-billion, employee-owned enterprise with some 70,000 employees—is the largest department store chain in the country. Annual profits are for the most part distributed among staff members, who are, in effect, the shareholders of the organization.[26]

Contemporary ownership structures are unduly wedded to nineteenth-century industrial capitalism, an era of capital scarcity and labor and natural resource abundance. In the twenty-first century, where financial capital is plentiful and skilled human capital and natural capital are relatively scarce, alternative ownership structures are flourishing and in a constant state of reinvention. The Grameen Bank/Groupe Danone (Bangladesh/France) joint venture exemplifies a new generation of blended business models that have at their core a social purpose—in this case, affordable nutrition for Bangladeshi children through the manufacture of fortified, low-cost yogurt. Meanwhile, trust- and foundation-controlled companies answer to a higher purpose defined by the nonprofit entity that holds controlling shares. Novo Nordisk (Denmark), GrupoNueva (Chile), and Tata Industries (India) fall into this category. In the case of Tata, 90 enterprises are controlled by family trusts and bound by the 140-year-old legacy of its founder to advance social capital.[27]

The Chinese model of state capitalism is yet another ownership model: the principal stakeholder typically is also the principal shareholder—China itself. The rise of Chinese enterprise worldwide in sectors such as mining, automobiles, and computer technology is moving the country well beyond the boundary of low-end manufactured goods into becoming a global player in both low and high technology. Here, enterprises are as much social and political instruments as they are economic engines, fortifying China's effort to ensure secure flows of minerals and food commodities from African and South American sources. Internally, such ownership is used as an instrument to foster social harmony, reduce inequalities between coast and interior communities, and accelerate the rise of the middle class—critical elements of the country's social agenda. Of course, the intimate connection between government and state enterprise has serious downside risks, whereby enterprises are beholden to political control aimed at preserving the one-party system. The scale of ecological degradation in China attests to the dark side of state capitalism, when political and economic interests trump environmental protection—to the long-term detriment of China's public health and ecological resilience.

All these examples illustrate the broad spectrum of ownership options in play on the world stage. To varying degrees, by design or by consequence, each provides an alternative to the dominant western model in terms of alignment with vision and principles of the new corporation depicted here. In an interconnected world confronting multiple ecological, economic, and social crises, ownership stands as one of the powerful pressure points for rethinking corporate designs such that they ingrain social mission in the conduct and culture of contemporary enterprises.

Capital. Whatever purpose or ownership structure is in place, corporations need financial capital to launch and sustain their operations. The access, sources, quantity, and conditions by which investment occurs play a pivotal role in enabling or impairing the transformation envisioned here.

Historically, capital markets have been at best indifferent to the long-term social consequences of investment practices. Among the tens of trillions of dollars in assets under management worldwide, only a small fraction is subject to any form of screen that aligns with the principles of the new corporation. In the United States, for example, recent estimates place the figure at $3.1 trillion, less than 15 percent of total assets under management.[28]

Globally, a number of the world's 100 stock exchanges are taking steps toward bringing sustainability into their listing requirements or other mechanisms for informing investors of the materiality of sustainability to their decisionmaking. These include the Shanghai Stock Exchange, BOVESPA (São Paulo), Johannesburg Stock Exchange, Deutsche Börse, Singapore Exchange, and the Stock Exchange of Thailand. At the same time, spurred largely by GRI, sustainability disclosure initiatives with capital markets as a key target continue to proliferate worldwide. By one estimate, some 142 standards or laws are in place, with two thirds of them mandatory. A portion of these are related to capital market activity, reinforcing the view that sustainability is moving, albeit slowly, from the margins into the mainstream of financial markets and government policy.[29]

Outside of the mainstream capital markets that traditionally serve public equities is a new asset class commonly referred to as "impact investing." Spurred by a coalition of 15 foundations seeking to harmonize their investment portfolios with their programmatic goals, together with a number of mainstream firms serving mission-oriented clients, impact investors seek opportunities in start-ups, funds, social enterprises, and projects for which social value is the central goal. Principally targeted at emerging markets and poor countries, the coalition represents $1.5 billion in assets aligned with impact investment performance metrics developed by an affiliated initiative, the Global Impact Investment Rating System. In the context of the new corporate designs, assets of this nature could become significant if they expand from their current modest level to, say, a trillion dollars or more in the next 5–10 years. In a similar vein, the Global Initiative for Sustainability Ratings seeks to reach beyond the relatively small mission-oriented investment community to drive environmental, social, and governance impacts into the mainstream capital markets.[30]

A future in which sustainability becomes seamlessly woven into the fabric of capital markets is a future with great promise for creating and expanding sustainability-oriented corporate forms. Realizing this potential at the magnitude and speed warranted by multiple global crises will require government actions in relation to securities regulation and stock exchange rules, public financing mechanisms in the form of national and state banks and targeted subsidies for new mission-oriented corporations, fiduciary regimes friendly to mission-driven investors, and capital gains taxes that privilege such investors. Examples of these types of actions are already in place. A vast scaling up and scaling out should be part of the Green Economy agenda of Rio+20 and beyond.

The Deutsche Börse in Frankfurt, Germany

Christoph F. Sieeermann

Governance. To accelerate corporate transformation, corporate governance—the structure of decisionmaking and accountability in an organization—must shift from a focus on shareholder accountability to stakeholder accountability. Governance structures and processes that operate with a broader, integrated view of the nature, sources, and equitable distribution of corporate wealth creation are the same structures and processes that best align with the desired corporate purpose and design. Central to transitioning to such a stakeholder paradigm of governance are the values, knowledge, and oversight of the board.[31]

Why, more than two decades after inception of the contemporary sustainability movement, does sustainability remain at the margins of the vast majority of TNCs? Shortcomings in corporate governance surely rank among the most powerful impediments. Owing to a combination of law, culture, and choice, the vast majority of corporate boards continue to embrace shareholder value as the ultimate measure of business success. Indeed, attachment to this pillar of governance has assumed something akin to the law of gravity—undeniable, uncontestable, and unchallengeable.[32] The primacy of capital interests permeates virtually every corporate national and international governance initiative.

U.N. Secretary-General Ban Ki-moon's indictment of the global economy is equally applicable to corporate governance: "The global economy needs more than a quick fix. It needs a fundamental fix. If we have learned anything from the financial crisis, it is that we must put an end to the unethical and irresponsible behavior and tyrannical demand for short-term profit."[33]

Transforming corporations to align with the principles of corporate design requires transformation of the boards that are responsible for organizations' long-term prosperity. The necessary values shift will take time and organizational will, legal and regulatory reform, a reorientation of global governance norms and standards, and, perhaps more decisively, pressure from stakeholders whose interests are underserved by current governance structures.

Illustrative actions that will contribute to this transformation are:
- requirements that all existing and future board members build their professional com-

petence through training in governance for sustainability;
- reconstitution of boards to include a mix of directors whose background and expertise mirror the full spectrum of the organization's key stakeholders;
- creation and independent funding of a Futures Council, a body that independently assesses the board's and the company's sustainability performance with reference to the interest of all legitimate stakeholders, including those of future generations;
- a requirement that directors hold management accountable for integrating sustainability into all business functions, and regular monitoring and assessment of such integration; and
- integration of executive compensation with the company's sustainability performance.

No single action is a panacea for transforming corporate governance. Indeed, it can be argued that the durability and effectiveness of any measure must be preceded by a deep reflection on the part of directors as to the purpose of the corporation, the role of the board in achieving such purpose, and the meaning of duty of loyalty and duty of care in the twenty-first century. As the Brazilian Institute of Corporate Governance has noted, the most advanced stage of a board's evolution is when sustainability initiatives are not presented to a board—instead, they emanate from it.[34]

The Road Ahead

A decade ago, Charles Handy, among the most incisive contemporary observers of the modern corporation, asked the most fundamental of all questions: What's a business for? His response remains relevant today: "To turn shareholders' needs into a purpose is to be guilty of a logical confusion, to mistake a necessary condition for a sufficient one. We need to eat to live; food is a necessary condition of life. But if we lived mainly to eat, making food a sufficient or sole purpose of life, we would become gross. The purpose of a business…is not to make a profit. Full stop. It is to make a profit so that business can do something more or better. That 'something' becomes the real justification for a business. Owners know this. Investors needn't care."[35]

The question of corporate purpose is central to the multitude of issues that define current debates over business-society relations. In Handy's terms, the "something" identified here is "the public interest"—which today translates into building a just and sustainable world. From this purpose flows the set of design principles, and from design principles flow the levers of change that are central to creating a generation of corporations that embed social mission in all aspects of their activities.

The seeds of such a transformation are discernible. Rio+20's focus on the Green Economy in the context of sustainable development and poverty eradication, one of two major conference themes, is but the latest evidence of this evolution.

A small but growing number of TNCs understand the imperative and the opportunity of revamping their business models to enhance both long-term value creation and competitiveness. Their leaders demonstrate the readiness to open a new chapter in business-society relations, to rewrite the social contract by bringing the commercial sector into the contract in a way that transparently, accountably, and democratically reinforces the historic bargain between citizens and their governments. Consumer goods TNC Unilever CEO Paul Polman, in advocating for the company's Sustainable Living Plan, argues: "Changes in policy will mean little if not accompanied by changes in behavior. That's why we need a different approach in business—a new model led by a generation of leaders with the mindset and the courage to tackle the [sustainability] challenges of the future." That "model" should be rooted in the vision, principles, and levers for change that are essential for repurposing and redesigning the corporation.[36]

In light of all these developments, the 2015 scenario described earlier does not seem as implausible as it might at first read. The building blocks, though scattered and imperfect, await bold and persuasive change agents from within and outside of the business community. Will history look back at Rio+20 as a moment when such agents boldly stepped forward with the will, passion, and determination to become the vanguard of transformation in the purpose and design of the corporation?

CHAPTER 8

A New Global Architecture for Sustainability Governance

Maria Ivanova

In June 1992, Rio de Janeiro welcomed what was the largest intergovernmental gathering on the environment up to that point in time. A total of 172 governments—108 of which were represented by the head of state or government—convened at the Rio Earth Summit to discuss our common future on the planet. Since then the world has become more globalized, urbanized, and interconnected. Geopolitical power balances have shifted as several countries have moved into the group of middle-income states. Flows of goods and services, capital and technology, information and labor have fueled a growing global population. Social and environmental challenges have also increased as the degradation of ecosystem services—the "dividend" that humanity receives from natural capital—narrows down development opportunities. The recent food and financial crises combined with the pressures of climate change demonstrate the inherently global nature of contemporary problems and the need for more effective global solutions.

As governments prepare to convene again in Rio in June 2012 for the United Nations Conference on Sustainable Development, the design of the institutional architecture for sustainability is one of the main agenda items. Academic and political debates have converged on the need to strengthen the global environmental governance system. Governments that had earlier argued against any reform as too costly and unnecessary have now called for a rethinking of the current system and strengthening of the institutional architecture. Other governments have renewed their calls for greatly improved environmental institutions.[1]

Surprisingly, the proposals for new institutional design closely resemble the ideas of the architects of global environmental governance in 1972, when governments established the first U.N. body for environmental matters—the U.N. Environment Programme (UNEP). This new institution was charged with keeping the global environment under review, offering policy options, catalyzing environmental awareness and action, coordinating environmental activities within the U.N. system, and developing national capacity. UNEP was conceived as an agile, swift, adaptable, and effective body that could leverage the strengths of the rest of the U.N. system to attain better environmental results.

Although much has changed in the last 40

Maria Ivanova is assistant professor of Global Governance at the McCormack Graduate School of Policy and Global Studies at the University of Massachusetts Boston.

years, the gist of the debate remains the same: What is the optimal design for the international architecture for sustainability? No one institutional structure, however, can guarantee effective resolution of environmental problems, especially at the global level. A systemic approach to understanding and re-envisioning the system is necessary—one that zeroes in on the root causes behind the challenges of the existing institutions and on levers for transformation. A close examination of recent history shows that the best and boldest ideas might not be new ideas at all but ideas whose time has finally come.

Environmental Summits: Platforms for New Architectural Designs

Environmental summits are an indelible feature of international politics that offer unique opportunities for leadership and social change. They have played a critical role in global environmental governance by galvanizing international attention to the environment, shaping the climate of ideas, and creating institutional architecture. Critics of the summits argue, however, that these large gatherings are irrelevant, wasteful, and even counterproductive because they convene and empower nation-states—an outdated governance unit. These days the influence of nongovernmental actors often surpasses that of many states, and state decisions are often symbolic.[2]

Yet the rapid growth of issues, actors, and agendas actually makes summits more relevant than ever. They provide critical junctures where states, civil society, and the private sector converge and can shape ideas and institutions for decades. The major U.N. environmental meetings—the 1972 Stockholm Conference, the 1992 Rio Earth Summit, and the 2002 World Summit on Sustainable Development in Johannesburg—have provided the strongest impetus and opportunities for institutional redesign.

In June 1972, at the Conference on the Human Environment, governments created UNEP. Twenty years later, at the Earth Summit, governments established the U.N. Commission on Sustainable Development and adopted conventions on climate change, biodiversity, and desertification. The summit also provided the impetus for the use of the new Global Environment Facility as the core funding mechanism for the environment. A number of principles, including public participation and access to justice, as well as common but differentiated responsibilities, also became part of the codebook guiding state behavior in international environmental affairs. The 2002 Johannesburg Summit stimulated a political debate on reform, elicited spirited calls for a World Environment Organization from world leaders such as French President Jacques Chirac, but in the end had no concrete outcomes for the international environmental architecture.[3]

The 2012 Rio Conference on Sustainable Development, known as Rio+20, is expected to make decisions on governance under the rubric "institutional framework for sustainable development." Even a decision for no reform will have enduring consequences and will shape the actions of the global community over the next 10–20 years.

Redesigning Global Environmental Governance

Three main reform options for architectural redesign in the environmental and sustainable development fields are currently in play. In the environmental arena, governments are discussing two options: enhance UNEP while retaining its current institutional status as a subsidiary body of the U.N. General Assembly or transform UNEP into a specialized agency of the United Nations. In the sustainable development field, governments are discussing the option of upgrading the Commission on Sustainable Development into a Sustainable Development Council. This chapter focuses on the environmental architecture negotiations, as those discussions have been going on for over a decade and governments are closer to consensus on this issue. The sustainable development debates are only beginning, however, and it is not yet possible to assess the main options and their consequences.[4]

The most recent intergovernmental reform negotiations on environmental governance, known as the Belgrade Process, started at the twenty-fifth session of UNEP's Governing Council in February 2009 and concluded with the Nairobi-Helsinki Outcome Document in 2010. (See Box 8–1.) The Belgrade Process convened a group of ministers and high-level representatives to outline both incremental and broader reform alternatives, seeking to inform the Rio+20 conference. The process capitalized on informal consultations on international environmental governance that took place within the U.N. General Assembly from 2006 to 2008.[5]

A set of key questions shaped that debate: How can the environmental work of existing institutions be made more effective? How can coordination and cooperation within the U.N. system be improved in order to overcome present weaknesses and improve the U.N. response to environmental challenges? What is the simplest, most economical, reliable kind of

---

Box 8–1. The Nairobi-Helsinki Outcome

Functional responses suggested in Nairobi-Helsinki Outcome document produced by the Consultative Group of Ministers or High-Level Representatives:

- Strengthen the science-policy interface with the full and meaningful participation of developing countries; meet the science-policy capacity needs of developing countries and countries with economies in transition; and build on existing international environmental assessments, scientific panels and information networks.
- Develop a system-wide strategy for environment in the United Nations system to increase the effectiveness, efficiency and coherence of the United Nations system and in that way contribute to strengthening the environmental pillar of sustainable development.
- Encourage synergies between compatible multilateral environmental agreements and identify guiding elements for realizing such synergies while respecting the autonomy of the conferences of the parties.
- Create a stronger link between global environmental policy making and financing aimed at widening and deepening the funding base for environment.
- Develop a system-wide capacity-building framework for the environment to ensure a responsive and cohesive approach to meeting country needs, taking into account the Bali Strategic Plan for Technology Support and Capacity-Building.
- Continue to strengthen strategic engagement at the regional level by further increasing the capacity of UNEP regional offices to be more responsive to country environmental needs.

Institutional form options suggested in Nairobi-Helsinki Outcome document:

- Enhance UNEP.
- Establish a new umbrella organization for sustainable development.
- Establish a specialized agency such as a world environment organization.
- Reform the United Nations Economic and Social Council and the United Nations Commission on Sustainable Development.
- Enhance institutional reforms and streamline existing structures.

Source: See endnote 5.

---

organizational setup that will get the job done? By 2008, no consensus had emerged among the U.N. missions in New York on answers to these questions, and UNEP's Governing Council was asked to take up the process with the help of environment ministers.[6]

The questions that posed such difficulty for governments in the twenty-first century were not new. In fact, governments had found answers to them almost 40 years earlier. In 1970, confronted with a relatively new set of global environmental problems, U.N. member states initiated consultations on the design of the international environmental architecture. Deliberations lasted for two years and were guided by the principle of "form follows function." The answer was the creation of UNEP in December 1972 as a subsidiary body of the U.N. General Assembly.[7]

UNEP was set up with a professional Secretariat, an Executive Director, and a 58-member Governing Council to promote international cooperation, provide policy guidance on environmental action within the U.N. system, and make recommendations for action by governments and international agencies. Governments also established an Environment Fund to support monitoring, research, and technical assistance and an Environment Coordination Board to coordinate information exchange in the U.N. system, bring together information from different sectoral and regional networks, and provide a coherent perspective on major environmental problems.[8]

A rich intellectual and political debate preceded UNEP's creation. The debate took place in a highly politicized environment, as the cold war was at its height, and many developing countries, having only recently gained independence, were seeking their rightful place at the international negotiations table. In the context of post-colonialism, environment and development were pitted against each other. In the early 1970s, it took significant effort on the part of the Stockholm Conference Secretariat

and leadership to engage developing countries constructively.

Even though no international organization had an explicit environmental mandate, the institutional landscape was not empty. The United Nations and its affiliates had dedicated resources to environmental protection and research for several decades, but in a piecemeal fashion that did not coordinate their activities with each other or with national partners. It was clear therefore that communication, coordination, and collaboration were paramount. Governments recognized that the separate environmental activities in the system should be brought together into the common framework of one United Nations environmental program but be carried out by all relevant agencies. A new U.N. body would have an overview of all problems and all U.N. activities and might make the United Nations as a whole more environmentally responsible and constructive.[9]

Overcoming ideological and political differences, the 113 governments in attendance at the Stockholm Conference agreed to create UNEP. Much of the rationale and design of this new body hold significant value both as explanatory factors and as a strategic plan for moving forward beyond Rio+20. Then, as now, governments deliberated on the institutional functions, form, and financing. Their solutions were well thought out, justified, and visionary.

Institutional Options in the U.N. System. Discussions about the future environmental body focused first on whether it should be within or outside the U.N. system. Prominent thinkers such as George Kennan argued that the urgent need for action on environmental problems made it imperative for any organizational arrangements to be outside the United Nations. With over 130 member states "deeply divided by national, racial, and ideological antagonisms and differing greatly in their perception of environmental problems and in the ability to contribute to their solution," as Richard Gardner, advisor to the Stockholm Conference Secretariat, put it, the United Nations was close to incapacitated. Furthermore, it faced a precarious financial situation, the quality of its staff was uneven, and the largely autonomous specialized agencies had hobbled effective collective action.[10]

Proposals emerged to create an environmental organization outside of the United Nations, limit its membership to industrial countries responsible for the pollution problems at hand, and endow it with real enforcement powers, "real teeth." Analysts noted that governments were not ready to cede power for environmental policymaking to an all-powerful supranational body with legislative and enforcement powers, but they recognized the need for cooperation. The United Nations was the only viable forum for international cooperation. Its membership was near universal, granting any environmental action legitimacy in both the industrial and the developing world.[11]

U.N. specialized agencies were already engaged in environmental work from a number of specific perspectives and could use a common outlook on the global environmental situation. In addition, proponents of a U.N. environmental body like Richard Gardner argued that "at a time when the United Nations is undergoing a severe crisis of confidence a success in the field of the environment could bring the organization an increased measure of public support."[12]

Subsequently, governments had to decide on the particular institutional form. When creating a new international organization within the U.N. system, there are only a few options. The most prominent are an autonomous specialized agency, a subsidiary body within the U.N. General Assembly (and the Economic and Social Council), and a unit within the U.N. Secretariat.[13]

Subsidiary bodies are entities created under Article 22 of the U.N. Charter to address emerging problems and issues in international economic, social, and humanitarian fields. They can have many different formal designations—programs, funds, boards, committees, or commissions—and governance structures. Their membership is usually geographically representative and their financial contributions voluntary. Some of their funding, however, comes from the U.N. regular budget, and they benefit from U.N. administrative services. Subsidiary bodies work directly through the United Nations, which also gives them the authority to play a leadership and coordinating role within the system.[14]

U.N. specialized agencies, in contrast, are autonomous organizations set up independently and linked to the United Nations through special agreements. Governments establish specialized agencies through treaties. They have universal membership—that is, any state can join as a member if they ratify the constitutive treaty. Their budgets include mandatory financial contributions assessed for member states according to a particular scale, and they do not receive any funding from the U.N. regular budget. Most specialized agencies were created in the years immediately after World War II to deal with discrete issues such as food and agriculture, health, civil aviation, and telecommunications.[15]

The third option is units within the U.N. Secretariat, which have discrete responsibilities in an issue area or overarching coordination functions. For example, the Office for Coordination of Humanitarian Affairs (OCHA) is the unit within the U.N. Secretariat responsible for bringing together humanitarian actors to ensure a coherent response to emergencies. OCHA also creates an overarching response framework, which allows for coordinated contributions from each actor.

In 1972, governments considered all three options as potential models for the new environmental entity. Forty years later governments are deliberating once again on two of them: transforming UNEP into a specialized agency or retaining it as a subsidiary body but supporting other ways of changing course, including improving financing, introducing universal membership, and creating an Executive Board. Before deciding on changing UNEP's institutional form, it is important to consider why UNEP was not created as a specialized agency in the first place.

Why UNEP is Not a Specialized Agency. Governments and scholars considered carefully the option of creating a specialized agency for the environment when designing the original institutional architecture 40 years ago. For a number of reasons, they deemed specialized agency status inappropriate for the functions they outlined.

First, a new specialized agency would need to assume a wide range of functions already performed by existing agencies. Such a transfer of functions would be difficult to define and execute. In the chemicals regime, for example, the World Health Organization (WHO) is most likely to be concerned with how chemicals affect human health, the International Labour Organization is concerned with protecting the rights of workers who interact with chemicals, the International Maritime Organization works to prevent chemical waste from entering the ocean, and the United Nations Institute for Training and Research provides training to developing nations in reducing the use of persistent organic pollutants. None of these functions could be or should be extracted from existing organizations. The scope of work for a new specialized agency therefore would be difficult to define short of an all-inclusive mandate.

Second, a new specialized agency for the environment would join the ranks as only one of many existing organizations with activities in the same sphere. Placed at the same level as organizations with longer traditions and well-

In 2009, UNEP's five executive directors: Achim Steiner, Maurice Strong, Mostafa Tolba, Elizabeth Dowdeswell, and Klaus Töpfer

Satishkumar Belliethathan

established relations with national and international bureaucracies, the new agency would have difficulty exercising its catalytic and coordinative functions. As David Wightman, advisor to the Stockholm Conference Secretariat, emphasized, "it would quickly find itself involved with them in jurisdictional disputes which could only be resolved at a higher level [and] would simply compound all the present jurisdictional difficulties of making the U.N. system function as a system in the true sense of the word."[16]

Third, U.N. agencies were not seen as highly effective international bodies. Governments considered them unnecessarily hierarchical, bureaucratic, and cumbersome. The fact that specialized agencies could only be chartered by a multiyear treaty process was also discouraging. Finally, as a government delegate to the third Preparatory Committee for the Stockholm Conference noted, it seemed "necessary to avoid imposing rigid institutional structures which would be rendered obsolete after a few years by rapid scientific and technological advances." The planners, in other words, feared relegating the entire integrative concept of the "environment" to one agency that could be isolated, marginalized, and unable to perform a catalytic and coordinative role.[17]

Since coordination was a critical function, proposals for the new environmental body included the design of a unit "placed at the highest possible level in the United Nations administrative structure, i.e., in the Office of the Secretary-General," as suggested by the United States. A strong executive for environmental affairs would direct the unit. This high-profile officer would oversee disbursements from a special fund to support activities conducted by other organizations and promote collaboration in the U.N. system. Maurice Strong, the Secretary-General of the 1972 Stockholm Conference, articulated his vision for the new environmental entity in a lecture in 1971. (See Box 8–2.) In December 1972 he became UNEP's first Executive Director.[18]

In the end, governments decided to create a new environmental entity as a subsidiary body of the U.N. General Assembly based on the following rationale. First, there were recent precedents of creating subsidiary bodies of the General Assembly with an autonomous status, including the U.N. Conference on Trade and Development in 1964, the U.N. Development Programme (UNDP) in 1965, and the U.N. Industrial Development Organization in 1966. Similar to these U.N. bodies, the new environmental entity would be directly responsible to the General Assembly but would possess its own governing body, take independent initiative and action, and—unlike a specialized agency—derive part of its funding from the regular budget of the United Nations.

Second, the subsidiary body status was considered advantageous as it would allow the new body to work within the U.N. system and

---

Box 8–2. Maurice Strong's Original Vision for UNEP

What is needed to deal with the task of improving the global environment is not a specialized agency but a policy evaluation and review mechanism, which can become the institutional center or brain of the environmental network. It might be charged with the responsibility of (a) maintaining a global review of environmental trends, policies and actions; (b) determining important issues which should be brought to the attention of governments and outlining policy options; and (c) identifying and filling gaps in knowledge and in the performance of organizations carrying out agreed international measures for environmental control.

This body would have to be sufficiently competent, both politically and technically, to give it a high degree of credibility and influence with both the governments and other organizations in the international system. It would have to have access to the world's best scientific and professional resources in evaluating the information, which would be valuable to it through the world monitoring networks operated by other agencies, both national and international.

If it were to be an effective instrument for coordinating and rationalizing environmental activities throughout the international system it would not undertake operational functions in which it would compete with the organizations it must influence. It should, however, exercise sufficient influence on the environmental activities of the agencies. This function would be strengthened if it were to be allied to a world environment fund, which would permit central funding of at least some aspects of the international environmental activities such as research and technical assistance.

—*Maurice Strong, 1971*
Source: See endnote 18.

grant it direct access to the highest and nearly universal political organ. Since the specific purpose was to bring together the different strands of environmental work in the U.N. system and provide a center of gravity for environmental affairs, direct association with the General Assembly was considered a significant benefit both politically and operationally. Politically, as David Wightman wrote, it would "ensure that environmental issues received significant and decisive political attention." Operationally, he argued, it would avoid "the repetitive process of reporting to some higher level body which the existing decentralized structure of the United Nations system involves."[19]

Association with the U.N. General Assembly had its downsides, however. The docket was already full and another subsidiary body was unlikely to gain real attention. Moreover, it ran the risk of exposure to political concerns.

UNEP's founders acknowledged that environmental problems did not fit within the traditional boundaries of the nation-state or the expertise of any single existing organization, it is important to note, and they emphasized that the key functions of the new entity would be to catalyze cooperation, encourage synergy among existing agencies, and bring the system together into a whole that was greater than the sum of its parts. They expected the new body would acquire the authority to play a leadership role in the U.N. system. Ultimately, the architects of the 1970s designed UNEP to be a catalyst, or as Gordon Harrison of the Ford Foundation put it, "a pinch of silver to energize mighty reactions."[20]

Environmental Chain Reactions in the U.N. System: Authority, Resources, Connectivity

The mighty reactions that the original architects envisioned in the U.N. system did not take place. Inadequate authority, resources, and connectivity hampered the force and speed of the chain reactions that UNEP was supposed to generate.

Authority. Authority derives from power given by the state or from expertise in an issue area. An effective institution would thus be both "in" authority—possess the legal mandate in a particular area—and "an" authority, commanding the necessary knowledge in the area. While UNEP had a formal coordinating mandate, the specialized agencies—much larger in terms of staff, resources, and infrastructure—questioned UNEP's expertise and ability to serve as a coordinating body and the center of a global environmental network.[21]

Formal relations between UNEP and the specialized agencies developed slowly as the agencies carefully guarded their turf in the 1970s and viewed the new program with suspicion. UNESCO, for example, considered itself as having been "already in ecological long pants when UNEP was born and need[ing] no ideological help," as Gordon Harrison put it. The International Atomic Energy Agency vigorously resisted any attempts from UNEP to launch environmental reviews of nuclear energy under neutral auspices. At WHO, "staff dubbed UNEP the United Nations Everything Programme and viewed any suggestions from UNEP as a presumptive attack on their record and competence," noted Harrison. Gradually, however, relations with the specialized agencies evolved into more collaborative endeavors, but they have been challenged by the lack of regular contact and communication with UNEP's Nairobi headquarters.[22]

Over time, environmental activities in the U.N. system and beyond burgeoned. As countries around the world began creating environmental ministries, existing agencies—intergovernmental and nongovernmental alike—started adding environmental activities to their work programs. And as new environmental issues gained political momentum, UNEP often facilitated the creation of new institutional mechanisms—multilateral agreements on ozone, biodiversity, chemicals, desertification, climate change, and so on. But without a visible and authoritative center of gravity in the international system, this proliferation of treaties and agreements has confounded and burdened national administrations.

Even when environmental activities come backed up by financial resources, the multitude of actors compromises effectiveness. For example, the Organisation for Economic Co-operation and Development estimates that in the 153 countries that receive official development assistance, there are 1,571 donor/recipient partnerships involving environmental financing. These all need to be maintained through policy dialogue, planning, coordination, accounting, and reporting. The many competing actors, funds, and initiatives often undermine the effectiveness of environmental financing and limit the results achieved.[23]

Resources. The voluntary character of UNEP's financial resources has drawn much criticism. Scholars and policymakers contend that voluntary financial contributions are the root cause for the small size of UNEP's budget. Specialized agencies, whose budgets include mandatory contributions, Frank Biermann of the Free University of Amsterdam argues, "can avail themselves of more resources and hence influence."[24]

Compared with most of its peers, UNEP's annual budget of $217 million is indeed small, especially in light of its ambitious mandate to "provide leadership and encourage partnership in caring for the environment." The voluntary nature of the contributions, however, does not single-handedly explain the low volume. The four largest annual budgets in the U.N. system for 2010, in excess of $3 billion, are those of subsidiary bodies that rely solely on voluntary funding—UNDP, the World Food Programme (WFP), UNICEF, and the U.N. Refugee Agency (UNHCR). (See Figure 8–1.) Even specialized agencies, whose core

budgets come from assessed contributions, depend heavily on voluntary financing. WHO, the Food and Agriculture Organization, and UNESCO all rely on voluntary funding for more than 50 percent of their budgets.[25]

Change in institutional form to a specialized agency might therefore not be the single most important factor that would increase UNEP's financial resources. Other features, such as mandate, size, and location are important determinants of the scale of financing. Institutions with clear operational mandates (UNDP, WFP, UNICEF, and UNHCR) hold significantly larger budgets than those with normative mandates (OCHA, the World Trade Organization (WTO), and UNEP). Larger staff size and multiple locations also require larger resources. What the financial data also show, however, is that institutional authority and influence do not derive from resources alone. The WTO, an oft-cited example of significant global influence, operates with a budget at the lower end of the spectrum. The ability to generate interest and commitment to an area of work and thus secure the requisite financial resources is a critical attribute for any U.N. entity.

Connectivity. Any institution needs to be able to link in a timely manner to different constituencies and actors through various means. UNEP's explicit mandate to catalyze environmental action in the U.N. system and to review and coordinate the environmental activities of U.N. agencies demands sustained, collaborative interactions with these entities. In the 1970s and 1980s, the modern information and communication technologies that enable interactions today were simply not available. And with headquarters in Nairobi, UNEP was geographically removed from other U.N. centers. The lack of speedy and convenient transportation options and inadequate telecommunications compromised UNEP's communication and coordination abilities and isolated it from its constituency. Without constant and close contact with UNEP and with increasing pressure to integrate environmental considerations into their own work, U.N. agencies began developing their own independent environmental agendas.[26]

The location in Nairobi had a deep impact on UNEP as an organization. Genuinely committed to addressing environmental challenges, UNEP staff witnessed the pressures and impacts of environmental degradation in the developing world. It is no surprise, then, that there was some push for greater engagement on the ground, with concrete projects and initiatives from staff members—despite the fact that it went against the core normative mandate of the organization. Moreover, the location in Nairobi increases UNEP's visibility in the developing world, and demands from these countries for engagement on the ground and support for implementation of environmental commitments are only natural. This pressure to engage in more operational activities—both from staff and from some member states—presents a challenge to UNEP's identity as only a normative U.N. body.

UNEP explicitly acknowledges that its mission is to serve as the voice for the environment within the U.N. system, to be an advocate, educator, catalyst, and facilitator. Reliable connectivity is therefore critical to UNEP's ability to engage with the global public. UNEP's appearance in the media—print, electronic, and social—is limited. UNEP generally makes it into the news when it issues reports. While this illustrates one of the organization's strengths—setting the environmental agenda through research and reports—it also demonstrates that UNEP has few direct links to the media. UNEP experts do not command a strong presence in public discussions in the

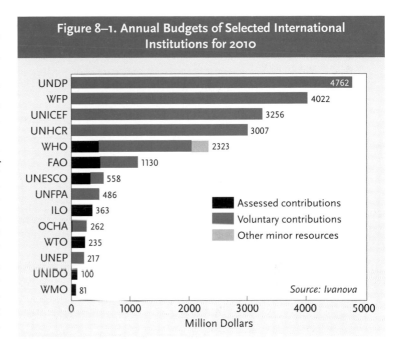

Figure 8–1. Annual Budgets of Selected International Institutions for 2010

Source: Ivanova

media and thus are not shaping public opinion through quotes in newspapers, interviews, and opinion editorials. The assumption might be that UNEP's work speaks for itself and will be heard when and where necessary. Yet in a world "where news and even revolutions are propagated in real time, 140 characters at a time," Khatchig Mouradian of the University of Massachusetts Boston notes, "it is agility and dynamism in the media that wins the day." Greater media presence is also a factor in increasing authority and could facilitate a catalytic and coordinative role in a system with multiple actors.[27]

Alternative Visions: Shared Governance, Centralized Responsibility

Environmental problems are ever evolving and continuously demanding different constellations of abilities. Moreover, they are almost always by-products or consequences of human activities, making it virtually impossible to deal with the environmental agenda separately. No one institutional design, therefore, is likely to provide the ultimate environmental architecture. A shared governance approach where authority is delegated to the most appropriate entity could ensure that complex problems receive adequate, multifaceted attention and treatment. In addition, a shared division of labor needs to be complemented by a clear line of responsibility and a strong, high-level executive who would ultimately be accountable for results or lack of them and be empowered to change course.

UNEP's founders documented clearly how they understood global environmental problems and how they thought the U.N. system could best address them. The new U.N. body they designed was to be small, highly visible, and deeply integrated with the rest of the U.N. system. It was expected to serve as the institutional "center" or "brain" in the international environmental system, and act as a policy evaluation and review mechanism. UNEP was meant to gather, evaluate, and issue up-to-date, accurate information about environmental trends, as well as about the policy performance of states and international organizations as they set and worked toward environmental goals.[28]

Furthermore, UNEP represented an attempt to institutionalize the integrative concept of "the environment" across existing U.N. agencies. Its designers acknowledged the activities of existing agencies and took pains to coordinate among them rather than set off competitions over jurisdiction, authority, and funding. In essence, Peter Stone, communications advisor to Maurice Strong, observed, UNEP was designed as a "virile, flexible instrument, which was not only going to try to save the world but to revitalize the UN as well."[29]

Today, sustainable development or sustainability is the integrative concept that governments seek to institutionalize within the U.N. system. Sustainability incorporates environ-

mental, economic, and social concerns but is ultimately based on a simple principle: everything humans need for survival and well-being depends, directly or indirectly, on the natural environment. The institutional landscape for sustainability is even more diverse than that for the environment, and reform of the institutional framework could be complex and complicated. The original vision and design of UNEP, however, could offer a blueprint for the contemporary reform process for both environment and sustainability.

In the environmental field, the most ambitious yet feasible reform option would be to empower UNEP to fulfill its well-thought-out visionary mandate. This could be done by improving authority, augmenting the financial base, and enhancing connectivity. Some of the actions could be initiated within UNEP, while others would require government decisions. (See Boxes 8–3 and 8–4.)[30]

In the sustainability field, the U.N. system would benefit from a High Commissioner for Sustainability within the Office of the U.N. Secretary-General, building on the original idea for a strong executive for environmental affairs with broad terms of reference at the center of the U.N. system. The role of the High Commissioner would be threefold. First, the Commissioner would help design, implement, and maintain a collaborative framework allowing the organizations working on environmental, economic, and social issues to proactively address existing and emerging socioeconomic and environmental issues. Second, the Commissioner would lead the U.N. system in generating momentum and providing the necessary conditions for implementing international environmental and sustainable development goals. Third, the Commissioner would ensure that the public is engaged in the sustainability discourse and actively participates in identifying and solving high-priority problems.

Creating a High Commissioner for Sustainability would enhance UNEP's authority as it would explicitly emphasize the importance of environmental concerns as a foundation for sustainable development. It would also provide the U.N. system with advice and offices for dispute settlement, establish a global cross-sectoral monitoring system, and promote public awareness and education around sustainability and environment. The Office of the High Commissioner would build and facilitate direct communication channels among relevant stakeholders, reducing fragmentation between organizations, institutions, and nations. It would have the authority to bring greater coherence in the U.N. system through joint programming, joint budgets, coordinated priority setting, shared research, and concerted outreach. Such an office could be created with a General Assembly Resolution rather than an intergovernmental treaty. It would provide for synergy between the three pillars of sustainable development rather than compartmentalizing global concerns and would have a unique level of authority in the U.N. system.

Conclusion

The need for a strong, legitimate, and credible authority for the environment is undeniable, but the link between the creation of a specialized agency and such authority is at best unclear. Despite significant changes in the complexity of the environmental and political situation over the past 40 years, as noted earlier the fundamental architectural question has remained the same: What is the optimal design for the international architecture for

---

Box 8-3. Internal UNEP Actions to Enhance Authority, Financing, and Connectivity

Focus on Staff and Culture. The most direct way for UNEP to gain greater authority in the environmental field would be to retain and recruit first-rate experts in all of its divisions and empower them to speak, write, and deliver on behalf of the organization. An explicit and sustained movement toward a culture of discipline and responsibility is likely to lead to high-quality work products and foster trust in UNEP's expertise.

Create and Use a Scientific Advisory Body. UNEP's authority derives also from its close association with top-quality research outside its direct purview. A close working relationship with the international scientific community is critical to its ability to identify environmental problems and develop the necessary policy options. To maintain and develop such a relationship, UNEP could establish a standing panel of scientific advisors. This body, however, should be governed and operated independently by the scientific community, perhaps following at the international level the model of the National Academies of Sciences in many countries. The benefit from such an arrangement would be that the knowledge, skills, and authority of the world scientific community would be leveraged and that scholars and young people would be encouraged to engage with the United Nations. The Advisory Body should be fully interdisciplinary and have a small staff for research and administration. Possible functions include systematic environmental assessment, identification of priority areas for research and action, network creation and maintenance among scientific communities across countries, and enhancement of national scientific capacities across the world.

Strengthen and Use the Environmental Management Group. UNEP's coordination arm, the Environment Management Group, is the successor of the Environment Coordination Board of 1972. With 44 member organizations from the U.N. system, the Environment Man-
continued

Box 8–3. continued

agement Group provides the platform for meaningful coordination. Strengthening it with more top-quality staff, a clear mandate, a flexible organizational structure, and visionary leadership with adequate discretion and resources would be an important step toward creating a functioning and result-driven international environmental governance system.

Enhance UNEP's Presence in New York. Much of the political debate on global environmental affairs takes place at the United Nations headquarters in New York. UNEP's existing liaison office in New York could be more authoritative—with a director at the level of Assistant Secretary General and with a larger staff who could participate meaningfully in most discussions on environmental issues at the United Nations and other U.N. bodies headquartered in New York, as well as at country missions. Physical presence at negotiations and regular, high-quality inputs in intergovernmental and nongovernmental

discussions would help UNEP acquire the authority among its peers and constituency.

Consolidate Financial Accounting and Reporting. Comprehensive and clear financial reporting is critical to building and maintaining the confidence and trust of donors. UNEP expenditure reports should indicate spending in terms of mandated functions—capacity building, information, coordination—as well as by environmental issues so that member states and donors can understand how UNEP as a whole is spending its money and effort.

Make Connectivity a Priority. While communications infrastructure and technology have been significantly improved, UNEP's presence outside its compound in Nairobi is limited. UNEP needs to engage purposefully, constructively, and systematically with constituencies at all levels of governance, reach out to universities as allies, and develop a sustained presence in both conventional and social media.

sustainability? Moreover, despite significant change in the geopolitical context, in the scale and scope of the environmental agenda, and in the urgency for creating global collective decisions, the fundamental vision, functions, and form that the original architects of the system devised remain valid today.

UNEP's designers demonstrated exceptional insight into how to direct the myriad institutions within the U.N. system toward coherent environmental action. As governments contemplate how to enhance UNEP and deliberate on transforming it from a sub-

sidiary body to a specialized agency, it is important to understand the powers that UNEP already has, the successes and challenges it has faced over the years, and the root causes of any obstacles and constraints. Just giving UNEP a new name—whether it be a World Environment Organization or a United Nations Environment Organization—would be grossly insufficient for enabling it to deliver on its mandate. Changing some of the key internal and external pressure points might lead to more effective and long-lasting results.

Box 8-4. Government Actions to Enhance UNEP

Expand UNEP's Governing Council Membership. Universal membership in UNEP's Governing Council could enhance the organization's legitimacy vis-à-vis states and the U.N. system because all governments will be members. It could also enhance UNEP's authority with regard to multilateral environmental agreements, many of which have near-universal membership. Universal membership, however, should be considered more broadly than just expanding the representation to all nation-states. Creating new and innovative mechanisms for engagement of civil society, the private sector, and academic institutions will be imperative to effective global problem solving.

Create an Executive Board. Currently, UNEP's Governing Council/Global Ministerial Environmental Forum (GC/GMEF) performs both of the organization's governance functions: providing leadership on international environmental governance and overseeing UNEP's program and budget. Performing both roles leads to circumscribed leadership and circular decisionmaking, in which programs and budget, rather than global needs, drive priorities and strategies. A global leadership role requires a large and inclusive structure like the GC/GMEF to review global issues, assess needs and identify gaps, identify priorities, and develop strategies to address them. The internal oversight role is best performed by a smaller body with greater discipline and focus on the program of work, budget, manage-

ment oversight, and program evaluation. An Executive Board of no more than 20 members, with representatives of both member states and civil society, could perform this role. Such an Executive Board will be critical in conjunction with universal membership.

Review the Need for Implementation Mandate. Analysts and policymakers have identified an implementation gap in international environmental governance. While many international institutions dictate policy and even provide incentives for implementation, there is no clear line of responsibility and accountability for implementation of multilateral environmental agreements or other internationally agreed goals. An independent external review of existing and necessary roles and responsibilities for implementing the myriad international environmental agreements would help clarify the mandates of other U.N. agencies and programs, reveal their comparative advantage, and provide a vision for reduced competition and a productive division of labor.

Allow for Some Assessed Financial Contributions. Assessed contributions may not lead to a greater overall budget but they are likely to bring greater stability and predictability of financial resources. A transition from a voluntary to a mixed financial contributions model might provide the necessary certainty for a core budget and the opportunity to be entrepreneurial and raise program resources.

A POLICY TOOLBOX
FOR BUILDING
SUSTAINABLE PROSPERITY

In moving toward sustainable prosperity, human society will need to draw on a number of essential strategies that, in order to succeed, will require the active participation of policymakers, business leaders, and civil society. The second part of *State of the World 2012* consists of nine short chapters that provide concrete recommendations on some of the policies needed to build a sustainable and prosperous global economy.

First and foremost, crucial in the pursuit of prosperity will be stabilizing the human population, as Robert Engelman of Worldwatch Institute explains. Engelman offers nine strategies for stabilizing world population numbers more quickly, including increasing access to family planning and education and convincing policymakers to make population issues a priority. This will make for a less crowded world, reducing ecological pressures and opening up new development opportunities.

It will also be essential to stabilize animal populations—particularly the 60 billion livestock animals that now supply the world's meat, eggs, and dairy products. Factory farming, as Mia MacDonald of Brighter Green describes, brings along with it a variety of ecological and social problems that will need to be remedied for the well-being of people, animals, and the planet. As Erik Assadourian of Worldwatch Institute points out in a Box, a different category of domesticated animals—the growing popula-

tions of pets—brings its own ecological burdens, as does farmed fish, a problem discussed in a Box by Trine S. Jensen and Eirini Glyki of Worldwatch Institute Europe.

Monique Mikhail of Oxfam examines the need to shift agricultural systems more broadly in order to produce enough food to feed everyone in a way that is sustainable, equitable, and resilient. She provides a number of solutions based on local sociopolitical and agroecological realities. Increasing gender equity, investing in small-scale food producers, and treating agricultural lands as diverse ecosystems rather than monocrop deserts will all be essential for growing a sustainable future.

Bo Normander of Worldwatch Institute Europe explores biological diversity and the significant commitment needed to avoid the sixth mass extinction on planet Earth. Governments will need to take an active role in combating not just climate change but habitat loss, including better protection of the world's oceans. Especially intriguing is his exploration of the untapped opportunities of urban areas to enrich biodiversity—an effort that can be driven by small-scale urban farmers around the world.

Ida Kubiszewski and Robert Costanza of the Institute for Sustainable Solutions at Portland State University follow this discussion with an exploration of the importance of ecosystem services for sustainable prosperity and how to better value these. As their chapter details, current economic considerations often ignore the contributions made by essential ecosystem services. There are, however, efforts to remedy this—creating common asset trusts, providing payments for ecosystem services, and so on. Governments will need to accelerate efforts to ensure that the true worth of ecosystems is recognized, before they and their essential services are lost entirely.

It is also critical that humans address their own infrastructural and institutional systems, as more of the world's people live in urban settings. Kaarin Taipale of the Center for Knowledge and Innovation Research of the Aalto University School of Economics discusses the steps needed to upgrade buildings from "sort of green" to truly sustainable. National and local governments will need to use a mix of policy carrots, sticks, and what Taipale calls "tambourines" to get builders to make every aspect of the building process—from production of construction materials to the refurbishing and eventual dismantling of old buildings—as sustainable as possible.

Helio Mattar of the Akatu Institute for Conscious Consumption portrays how the consumer culture has spread around the world. Failing to address the growing consumption levels of the consumer class—and those striving to join it—will make sustainable prosperity an impossible dream. Mattar offers a number of ways governments can help people reduce their consumption and make the consumption that does occur more sustainable. In accompanying Boxes, Dagny Tucker of Universitat Jaume I Castellon de la Plana expands this further with

a look at how community can substitute for consumption in achieving human well-being. And Yuichi Moriguchi of the University of Tokyo investigates Japan's efforts to create a circular material economy, which through cycling waste back into the raw material of new products can help make production and consumption more sustainable.

Essential in shifting consumption trends, and economic trends more broadly, will be shifting the role of business. Jorge Abrahão, Paulo Itacarambi, and Henrique Lian of the Ethos Institute advocate mobilizing businesses to build a green, inclusive, and responsible economy. Drawing on their efforts in Brazil, the chapter authors offer a variety of ways to engage the business community to play a more active role in creating sustainable prosperity—from the local all the way to the global.

In the final chapter in this section, Joseph Foti of the World Resources Institute looks at the important and often overlooked role of local governments in ensuring a high quality of life and a healthy environment for people. He discusses the opportunities for reducing environmental pollution and unsustainable development and for offering access to building blocks for sustainable prosperity—like public transportation and sanitation—when strong local governments lead and are supported by an active citizenry.

Together these nine brief articles and accompanying Boxes provide an abundance of concrete strategies for stopping humanity's slide into an ugly, unsustainable future and moving instead toward a true and lasting prosperity that can be shared by all.

—*Erik Assadourian*

C H A P T E R   9

Nine Population Strategies to Stop Short of 9 Billion

Robert Engelman

The demographers who calculate the future size of world population are not so much wrong as misunderstood. Humanity may indeed grow to 9 billion people by the middle of this century from 7 billion today and then stop increasing sometime in the twenty-second century around 10 billion. But this outcome is far from inevitable. It is neither an estimate nor a prediction but merely a projection—a conditional forecast of what will come about if current assumptions about declining human fertility and mortality prove true.[1]

No one, however, can be certain where birth or death rates will go in the coming years. (Migration rates are even less certain, but they only influence global population if birth and death rates change because people move.) And although policymakers and the news media rarely mention the possibility, societies can do a great deal to prompt an earlier peaking of world population at fewer than the "expected" 9 billion. Ending population growth would accelerate population aging, which means a rising median age for people in a country or the world. That could challenge societies economically as smaller proportions of a population are working and contributing to the retirement and health care benefits of a growing number of older, non-working people. Yet that is all but certain to be a manageable trade-off in return for longer lives in a less crowded and environmentally stressed world.

Ending Population Growth

The contribution that an end to population growth would make to environmentally sustainable prosperity is straightforward. The future of wealth and its distribution will be closely linked to the future of the global climate, the health of nature, and the availability of key natural resources. Since all descendents of today's low-income, low-consumption populations will anticipate and should expect consumption-boosting economic development, a lower future population would mean less pressure on climate, environment, and natural resources by future generations. It is a scenario without a downside for global well-being.

No ethical person would want an early end to population growth through rising death rates, though such an outcome cannot be ruled out given current trends in climate change, food production, and energy supplies. Nor is

---

Robert Engelman is president of Worldwatch Institute.

there now, or for the foreseeable future, significant public support for policies that would impose reproductive limits on couples and individuals. Abundant experience from around the world, however, demonstrates clearly how to reduce birth rates significantly through policies that not only respect the reproductive aspirations of parents and would-be parents but support a healthy, educated, and economically active populace—especially of women and girls. This chapter describes nine strategies that collectively would be likely to end human population growth before mid-century at a level below 9 billion. (See Figures 9–1 and 9–2 for profiles of world population growth since 1970.) Most of the policies are relatively inexpensive to put in place and implement, although some are culturally and hence politically sensitive in many or most countries.[2]

Assure Universal Access to a Range of Safe and Effective Contraceptive Options for Both Sexes. Since the early 1960s the use of contraception has increased markedly, with most women of reproductive age around the world using it. This increasing contraceptive prevalence has closely tracked a comparable and opposite decrease in average family size worldwide. Nevertheless, more than 40 percent of all pregnancies are unintended, and a conservatively estimated 215 million women in developing countries alone are hoping to avoid pregnancy but not using effective contraception. Although physical access to contraception does not guarantee that all reproductive-age people will use it, it is essential for personal fertility control (especially where there is little or no access to safe abortion). Demographic evidence is growing that if all women could time their pregnancies according to their own desires, total global fertility would fall below effective replacement levels (two-plus-a-fraction children per woman), putting population on a trajectory toward a peak and gradual decline before 2050.[3]

An estimated $24.6 billion a year would pay for the family planning and related maternal and child health services needed to ensure that all sexually active women in developing countries who seek to avoid pregnancy could gain access to contraception. By comparison, the world spends approximately $42 billion on pet food each year. (See Box 9–1.) Satisfying the unmet need for contraception in industrial countries would presumably cost less (although no estimates of that are available), as most such countries have fairly well developed health systems that provide at least some level of reproductive services.[4]

Perhaps the dominant obstacle to making access to family planning universal is widespread ambiguity about human sexuality and the persistence of religious and cultural barriers to the principle that women, whether married or not, should be able to choose sexual expression without fear of unintended

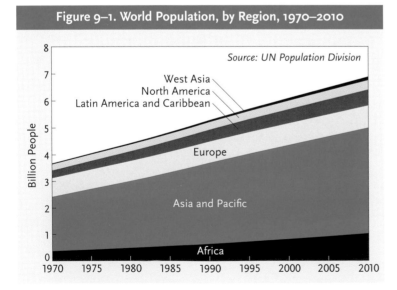

Figure 9–1. World Population, by Region, 1970–2010

Source: UN Population Division

West Asia
North America
Latin America and Caribbean
Europe
Asia and Pacific
Africa

pregnancy. Surveys indicate that the vast majority of Americans, at least, believe that women should be able to choose the timing and frequency of child-bearing by having access to contraception. Ensuring that all couples can make such choices will require much stronger public support in the face of ongoing opposition to family planning and marginal-ization of the links between women's reproductive choices, population dynamics, and social well-being.[5]

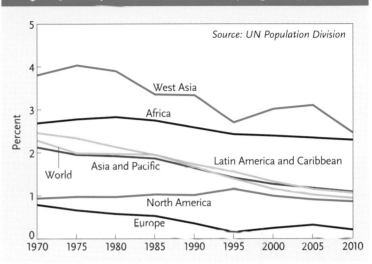

Figure 9–2. Population Growth Rates, by Region, 1970–2010

Guarantee Education through Secondary School for All, with a Particular Focus on Girls. Experts differ on whether contra-ceptive access or educational attainment more directly reduces fertility. In every culture sur-veyed, however, women who have completed at least some secondary school have fewer children on average, and have them later in their lives, than women who have less edu-cation. Surveying literature on this connec-tion, for example, Dina Abu-Ghaida and Stephan Klasen of the World Bank estimated in 2004 that with each year of completed secondary schooling, women's average fertility rates around the world are 0.3–0.5 children lower than those of women without that amount of schooling.[6]

Worldwide, according to calculations by demographers at the International Institute for Applied Systems Analysis, women with no schooling have an average of 4.5 children, whereas those with a few years of primary school have just 3. Women who complete one or two years of secondary school have an aver-age of 1.9 children—a figure that over time leads to a decreasing population. With one or two years of college, the average childbearing rate falls even further, to 1.7. Education

informs girls about healthy behavior and life options and hence motivates them to endeavor to postpone and minimize the frequency of childbearing so that they can more easily explore aspects of life beyond motherhood.[7]

As with the increasing use of contraception, global progress in educating girls is already impressive. As of 2010, more than three in five individuals age 15 or older—just over 3 bil-lion people—had finished at least some sec-ondary school during their lifetimes. This proportion has risen from 36 percent in 1970 and from 50 percent in 1990. Girls as well as boys have benefited from this improvement. Yet a "gender gap" between female and male edu-cational attainment remains, with the percent-age of girls in school consistently about 9 percent lower than the percentage of boys in school. And there appears to be a long way to go before most young women have effective access to a complete and adequate secondary-school education, especially in the least-devel-oped countries. These countries are generally the ones with the most stubbornly high fertil-ity. Investing in education—not just to bring children into schoolrooms but to improve the

Box 9–1. Environmental Impact of Pets

Along with the human population, another population has been growing rapidly around the world: pets. Today, the large population of dogs, cats, and other companion animals is having a serious impact on the world's environment.

In the United States, for example, there are now 61 million dogs and 76.5 million cats. Just in terms of food, a large dog uses 0.36 global hectares of resources per year, a small dog 0.18, and a cat 0.13 hectares. For comparison, a person in Bangladesh uses on average 0.6 hectares of resources a year in total—less than what two German Shepherds use in a year. Thus, in a conservative estimate, feeding American pets has as much of an environmental impact as the combined populations of Cuba and Haiti.

Many pets today also use more resources in the form of clothing, toys, and elaborate veterinarian care. A small percentage of pets even get treated to costly services like dog walkers, grooming salons, and private pet air travel service. One analysis finds that an American dog owner typically spends anywhere from $4,000 to $100,000 on a dog over its lifetime.

This is not just an American phenomenon. Pet ownership is a global phenomenon, with pet food alone costing $42 billion worldwide each year. The pet industry has worked hard to spread a culture of pet ownership around the world. Brazil has the world's second largest dog population at 30 million, along with 12 million cats. China has the third largest dog population (23 million dogs), and dog ownership is growing so fast that Shanghai passed a "one pet policy" in 2011 in reaction to such problems as dog bites and rabies.

Ultimately, shrinking the population of pets will have the same benefits as stabilizing the human population: it will free up more ecological space for development and for restoring Earth's systems. Several key strategies, if implemented, will help this process.

First, all pets that are not intended for breeding should be spayed and neutered early in their lives—common practice in some countries but not all. This will prevent unwanted pets as well as feral animal populations, which can damage bird populations and even threaten people. Adopting animals from shelters (and sterilizing them) instead of buying pets from breeders will also help.

Second, policymakers should recognize that pet ownership is a luxury and should make it costlier to own pets, perhaps through a steeper pet license fee or a tax on dog and cat food. Including the costs of ecological externalities in all products—including pet products—would increase the expense of pet ownership further.

Third, there should be better oversight of the pet industry, which has an industry strategy of "humanizing" pet populations so that people will seek out pets to fill companion gaps and spend more on them. Better regulation of marketing efforts may help curb pet populations and over time make pet ownership less normal.

Finally, pet owners (and children—the pet owners of tomorrow) should learn about the significant ecological costs of pets. This may curb some pet purchases and may also reduce excessive purchases for current pets—whether that is extra food (many pets are overweight due to overfeeding), clothing, fancy toys, pet spa treatments, and end-of-life medical care that is more sophisticated than many people in developing countries have access to. Over time, people may also shift to smaller pets, productive pets (like chickens or goats), or pets shared among a community.

—*Erik Assadourian*
Source: See endnote 4.

quality of their schooling—is among the rare "triple wins" that boost human well-being, economic development, and women's intentions and capacities to have fewer children later in their lives.[8]

Eradicate Gender Bias from Law, Economic Opportunity, Health, and Culture. While universal access to good contraceptive services and secondary school education in combination would reverse population growth, active efforts to foster legal, political, and economic gender equality would make contraceptive and educational access much easier to achieve and would hasten the reversal of growth. Women who are able to own, inherit, and manage property, to divorce their husbands, to obtain credit, and to participate in civic and political affairs on equal terms with men are more likely to postpone childbearing and reduce the number of their children compared with women lacking such rights and capacities. Indeed, a 2011 comparison of fertility rates with differentials between men's and women's political, economic, and health status demonstrated a significant correlation between high gender equality and lower rates of childbearing.[9]

Research indicates that a number of specific indictors of women's empowerment result in reduced or later childbearing. A study in northern Tanzania, for example, found that women with an equal say to their husbands in household matters preferred to have significantly fewer children than those who had to defer to their husbands' decisions. This is particularly important because men, free of the physical hazards and discomforts of childbearing and usually investing much less time than women do in childrearing, tend in most countries to want more children than their partners do.[10]

Demographic and health surveys over the past several decades for the U.S. Agency for International Development show that women in almost all developing countries express a desire for fewer children than they end up having, as well as fewer children than men want. The more children a woman has, the more likely she is to want fewer additional ones than her partner. How any specific indicator interacts with fertility intentions and outcomes remains unclear, but the broad connection of women's status and autonomy to later childbearing and smaller completed families adds to the reason to change laws and customs that institutionalize gender inequality.[11]

Offer Age-appropriate Sexuality Education for All Students. A major obstacle to the prevention of unintended pregnancy is ignorance by young people about how their bodies work, how to abstain from unwanted sex, how to prevent pregnancy when sexually active, and how important it is to respect the bodies and sexual intentions of others. Education in all these matters would further reduce unintended pregnancies and hence slow population growth. This can begin in age-appropriate ways almost as soon as schooling does. Questions about sex typically arise early in children's lives and require appropriate responses from the adults around them. Children are sometimes the victims of sexual harassment or violence and need to learn early in their lives how to recognize, protect themselves from, and report inappropriate sexual behavior.

Sexuality education differs significantly among countries and is absent from the curricula of many or most. In the United States, comprehensive sex education tends to stress the health and pregnancy-avoidance benefits of abstinence as well as the importance of contraception and safe sexual practices for those who choose not to be abstinent. U.S. data indicate that exposure to comprehensive programs tends to delay the initiation of sex and to increase the use of contraception among young people. Along with the other benefits provided, both of these trends would logically contribute to lower teen birth rates and probably lower completed fertility.[12]

End All Policies that Reward Parents Financially Based on the Number of Their Children. There is no reason to believe that pro-natalist government policies that reward couples financially for each additional birth have significantly raised total fertility rates in any country. Nonetheless, it seems logical that at least on the margin such policies do boost birth rates slightly. The policies may be as blatant as those in Russia and Singapore that directly pay couples for additional children. Or they may be couched as child care tax credits that reduce a parent's taxes for each additional child under 18 without limit, as in the United States. Such policies subsidize "super-replacement" fertility (rates well above two children per woman), contributing to populations larger than they would otherwise be.[13]

ernments can preserve and even increase tax and other financial benefits aimed at helping parents by linking these not to the number of children but to parenthood status itself. A set benefit to all parents would allow them to decide for themselves whether another child makes economic sense given that the benefit will not grow—just as the environment and its resources do not grow—with any addition to the family.[14]

Integrate Teaching about Population, Environment, and Development Relationships into School Curricula at Multiple Levels. Although environmental science education is now well established, especially at the university level, few school systems around the world include curricula that teach young people how human numbers, the natural environment, and human development interact.

Yet today's young people are very likely to spend most of their lives in densely populated human societies facing significant environmental and natural resource constraints. Without advocacy or propaganda, schools should help young people make well-informed choices about the impacts of their behavior, including childbearing, on the world in which they live.

In the United States, the organization Population Connection has an active education program that provides curricular material and training to teachers interested in awakening students of all ages to the dynamics and importance of population growth. It is not clear, however, how widespread the concept is in the United States or other countries. More education about human-environment interactions, including the influence of human numbers, nonetheless could become an important stimulus to a cultural transformation that can hasten an end to population growth.[15]

Father and son working together in Papua New Guinea

Where it is clear that women and couples are forgoing childbearing because of social discouragement (for example in the workplace) or a lack of acceptable child care options, governments can address these issues directly. In some northern European countries, for instance, fertility rates rebounded from very low levels after governments made paid leave mandatory for new parents of either sex. Gov-

Put Prices on Environmental Costs and Impacts. Governments need to move toward environmental pricing—including taxes, fees, rebates, and so on—for many reasons as soon as politically feasible. Among the benefits of carbon and other green taxes is their value in reminding parents that each human being, including a new one, has impacts on the environment. In a crowded world of constrained resources, these impacts should be accounted and paid for so that large environmental footprints face economic constraints. These constraints could be government-imposed, as in the case of carbon taxes or usage fees for waste removal services that are based on weight. Such environment-related governmental constraints on consumption are currently rare, however, and may not be feasible politically for some time. Free-market pricing may eventually play a similar role if the costs of food, energy, and various natural resources continue to rise due to scarcity and distribution challenges, as many analysts predict.

The rising financial costs of large families no doubt already discourage high fertility in countries where contraception is socially acceptable and readily available. If at some point governments opt to raise the costs of consumption that has negative impacts on the environment, couples and individuals will still be free to choose the timing and frequency of childbearing. Yet by translating into higher costs the impact of individuals, environmentally based pricing will tend to reduce fertility and birth rates as couples decide the cost of having an additional child is too high. This is hardly the reason to move toward environmental pricing, but it will be among its benefits.

Adjust to Population Aging Rather Than Trying to Delay It through Governmental Incentives or Programs Aimed at Boosting Childbearing. Higher proportions of older people in any population are a natural consequence of longer life spans and women's intentions to have fewer children, neither of which societies should want to reverse. The appropriate way to deal with population aging is to

make necessary social adjustments, increasing labor participation and mobilizing older people themselves to contribute to such adjustments, for instance, rather than urging or offering incentives to women to have more children than they think best.

Population aging is a short-term phenomenon that will pass before the end of this century, with impacts far less significant and long-lasting than ongoing population growth, a point policymakers need to understand better. Even if today's policymakers could boost population growth through higher birth rates or immigration, future policymakers would have to grapple with the problems of aging at some later time—when higher population density and its associated problems only make boosting population growth less attractive and feasible.[16]

Convince Leaders to Commit to Ending Population Growth through the Exercise of Human Rights and Human Development. Several decades ago, it was not unusual for presidents and prime ministers in industrial and developing countries to declare their own commitment to slowing the growth of population. Today, with twice as many people as were then alive seeking the good life, the need is more acute than ever for political leaders to find the courage to acknowledge the importance of ending population growth. For a variety of reasons, however, population has become a taboo topic in politics and in international affairs, though perhaps somewhat less so in the news media and in public discourse.

Speaking out on the importance of ending human population growth worldwide will be easier if leaders acquaint themselves with how the population field has evolved over the past few decades. They will then understand that human numbers are best addressed—in fact, can only be effectively and ethically addressed—by empowering women to become pregnant only when they themselves choose to do so. One irony of this is that slowing population growth needs to be seen not so much as the goal of some kind of crisis or emergency program—a vision that the public and politicians alike would find frightening—but merely as a recognized and lauded side benefit of a host of policies that improve the lives of women, men, and children. If, through the education strategies described here and a broader cultural transformation on the topic, more people recognize the value of an end to population growth, each of these policies will become more feasible and more effective in bringing about beneficial demographic and environmental change.

The Impact of the Nine Strategies

To some extent most of these policies already are moving forward, albeit sluggishly, in different countries around the world. Powerful forces—in some cases religious and cultural, in others economic—oppose them, however. Sadly, it may be years or decades before increasing environmental deterioration and resource shortages in an ever more crowded world arouse the public so much that people demand governmental action on root causes. A powerful momentum helps drive today's population growth. As long as many more people are in or approaching their childbearing years than are nearing the end of their lives, as is the case today, humanity will increase for some time even if families are quite small. It will take time for the smaller generations of

children to become parents themselves and produce even smaller generations as the larger, older generations pass on. The longer governments delay policies such as those described here, the more likely the world is to face large and denser populations or increases in death rates—or both.

If, by contrast, each of these policies somehow could be put in place quickly and were well supported by the public and policymakers, population momentum itself would be slowed significantly, through later and fewer pregnancies, than ever witnessed in recorded history. Few demographers have attempted to quantify the population impact of various interventions beyond family planning access and education for girls on fertility. But based on what is known and can be logically conjectured, it seems likely that putting most of these policies together would undermine even population momentum and produce a turn-around in population growth—with the significant social and environmental benefits such a dynamic would offer—earlier than most demographers believe is likely or even possible. World population might indeed stop growing well short of the 9 billion so many believe is inevitable. The fertility declines that could bring a population peak at around 8 billion before the middle of this century, with no increases in death rates, are not unimaginable. If this were to occur, a truly prosperous and sustainable global society would be one long stride closer than ever before.

CHAPTER 10

From Light Green to Sustainable Buildings

Kaarin Taipale

We live in the built environment, whether in cities or the countryside. The construction sector accounts for at least a third of all resource consumption globally, including 12 percent of all fresh water use. Some 25–40 percent of produced energy is consumed in the construction and operation of buildings, which accounts for approximately 30–40 percent of all carbon dioxide (CO_2) emissions. And 30–40 percent of solid waste comes from construction. In terms of the economy, the sector produces approximately 10 percent of gross world product, and buildings represent a massive share of public and private assets. As the recent economic collapses in different parts of the world have shown, the stability of financial markets is linked with the long-term value of real estate as collateral. In terms of employment, at the national level the sector generates 5–10 percent of jobs, including in management and maintenance.[1]

The world is already more urban than rural. And the rate of urbanization means that in 2030 about 1.4 billion more people will live in cities—with 1.3 billion of them being in developing countries. They will all need homes, services, and places to work—new buildings, in other words. In the coming years, there will be more construction on the globe than ever before. All these buildings have long-term impacts.[2]

"Green building" has become trendy, even if it so far affects only a small portion of the market. Yet the recent rush to market everything as "green" makes it difficult for consumers to know what can have real impact and what is mere "greenwash" by builders and those providing the finance. The challenge of policies on buildings is to go beyond cheap tricks to basic principles.

"Sustainable" buildings are not only "green."A new home with solar panels on the roof may look like an "eco-house," but how many cars does the family need to get to school, to work, and to do the shopping? Is there public transport? What kind of wood was used for the walls and how was it treated? How much energy does the house need for heating, cooling, hot water, and appliances? Is the energy produced by a utility that burns coal? Does the solar panel give enough elec-

---

Kaarin Taipale is a visiting researcher at the Center for Knowledge and Innovation Research of the Aalto University School of Economics in Finland. She is the former chair of the Marrakech Task Force on Sustainable Buildings and Construction.

tricity for the greenhouse, or more? Did the homeowner have to dine with local politicians to get a building permit in an area that is not zoned for housing? Did the construction workers pay their taxes, and were they insured? The list of inconvenient questions goes on.

All dimensions of sustainability have to be assessed, both when a building is constructed and during its entire lifespan until the day it is torn down or renovated. (See Table 10–1 for the range of issues to be considered.) The urgency of climate change mitigation has meant that energy consumption and CO_2 emissions have in recent years been discussed more than other aspects. But the complete picture is bigger. Economic sustainability counts both the initial investment in land, design, and construction and the cost of maintaining and operating the building. Social and societal sustainability cover issues such as availability of appropriate housing for all, fair trade of construction materials, transparency in tendering for contracts, and protection of cultural heritage. Sustain-

able construction also means decent jobs, for example in maintenance and renovation of buildings and infrastructure.[3]

Policies at Work: Sticks, Carrots, and Tambourines

Policies can control (restrictive regulations), motivate (incentives), or call for attention (awareness raising). Successful policy packages combine the three characteristics: sticks, carrots, and tambourines. (See Table 10–2.)

Land use and building acts and codes are typical sticks. They formulate mandatory requirements for building permits and minimum standards for construction materials like cement and elements like windows. Instead of prescriptive regulation, however, which describes an ideal solution, contemporary regulation outlines the expected performance, such as how long a structure will have to resist fire before it collapses. Use of certain materials may be prohibited, such as asbestos for health reasons or illegal timber to stop defor-

Table 10–1. Layperson's Checklist of Issues to Be

Life Cycle	Sustainability Issues	
Phase of the Cycle	Consumption of Natural Resources	Consumption of Financial Resources
Production of construction materials	Land	Initial investment
Selection of a construction site	Fresh water	Material vs. labor costs
Design (architectural, engineering, technical)	Nonrenewable energy sources	Bribery costs
Procurement of materials and construction works	Renewable energy sources	Operational costs, including water and energy
Construction	Wood	Maintenance costs
Maintenance of the building	Metals	Refurbishment costs
Refurbishment	Minerals	Long-term value of property
Reuse of buildings	Stone, gravel	Transport costs of construction materials
Recycling of construction materials		Waste management costs

Source: See endnote 3.

estation. Workplace safety requirements protect people from injuries. But regulations only matter if their compliance is enforced and there is no corruption.

Urban and regional plans are strong sticks. They can have detailed requirements regarding the size or even the materials of new buildings. They can ban construction in a greenfield location unless it has access to public transport, or they can ban construction in an area altogether. Historic buildings can be landmarked.

Carrots are incentives to do better than just meeting the minimum requirements. Some examples of these are subsidies, green mortgages, direct public investment, and taxation policies. The real estate tax can be reduced for a building with an improved energy performance, or the tax can be higher for polluting fuels than clean energy.

Noisy tambourines are awareness-raising tools to draw attention to the necessity of sustainable building and to inform people about the best traditional and contemporary solutions. Everyone involved in the long process

from making bricks to refurbishing existing neighborhoods needs guidance. Newsletters, websites, and publicity campaigns are typical tambourines. Global events like the Earth Hour of WWF remind people of the need to save energy. Car-free days organized by cities highlight the possibility of using public transport.

In Search of a "Best Policy"

Policies are by far the cheapest and most efficient means of promoting sustainability in construction. They have to cover at least four aspects: process, performance, sustainable infrastructure, and the use of resources. But it is hard to think of a stand-alone policy that would make or break sustainable construction. Instead, policies need to be packaged to reinforce each other.

They have to include measurable benchmarks and targets. For the long term, zero is a good number: net zero energy, zero carbon, zero waste—and zero tolerance of corruption. Net-zero-energy buildings produce over the

Considered in Sustainable Buildings and Construction		
	Potential Impacts	
Human Conditions	Negative Impacts	Positive Impacts and Co-benefits
Access to fresh water and sanitation	Disruption of ecosystems due to land use changes	Reduced consumption of nonrenewable resources
Access to clean energy	Pollution of air, soil, and water	Energy savings
Availability of public transport	Contribution to climate change	Clean water
Accessibility of services and amenities	Waste	Improved human health
Indoor air quality	Traffic congestion	Job creation
Decent housing	Noise	Workplace safety
Structural safety	Informal settlements	Transparent governance
Security in the community	Corruption	Saved financial resources
Cultural value of existing buildings	Poor return on investment	Buildings as collateral
Decent work		

Table 10–2. Examples of Policy Tools on Buildings

	Mandatory Requirements	Incentives	Awareness-Raising Tools
Process: Long-term thinking, life cycle approach	• Mandatory "maintenance diary," showing how the building has been serviced • Anti-corruption measures	• Subsidies for refurbishment • Construct & maintain contracts for long-term lease	• Public hearings about land use planning and building permits • Some evaluation systems • Voluntary checklists
Performance: "How well" instead of prescribing "how to do it"	• Minimum energy performance standards • Handicapped access requirements	• Reduction in real estate tax for extra energy efficiency • Introduction of sustainable public procurement criteria	• Awards for excellent buildings or developers • Local guidebooks, websites, Q&A sessions for builders
Sustainable Infrastructure: Access to basic services	• Land use plan that allows new construction only if public transport is available • National water legislation	• Feed-In Tariff for renewable energy • Cross-subsidized pricing and reliability of public transport	• Car-free days, when public transport is free for all • Declaration of access to water as a human right
Resource Use: Renewable or not? Polluting? Hazardous?	• Prohibited use of tropical wood or asbestos • High price for management of construction waste	• Pricing of water and energy (use more, pay more) • Research funding	• Save Energy Day! Earth Hour • "Energy consulting bus" • Water and energy metering in every household

course of a year as much energy from renewable sources as they consume. Zero carbon results from the use of something other than fossil fuels to produce energy, eventually supported by measures to capture and store carbon with reforestation or other methods. Zero waste has proved to be the most difficult target.[4]

Assessment tools are needed to support policy decisions. Sustainable public procurement criteria are powerful as both sticks and carrots when products, construction work, and buildings are purchased. The criteria can range from fair trade and decent work requirements to minimized embodied energy. The Japanese Top Runner approach is a famous policy example: public money can be spent only on the most energy-efficient product on the market, which then becomes the baseline of efficiency for all manufacturers of such products. In an ideal world, sustainability criteria would be used for all financing of construction. Some national housing funding institutions already do that, such as the Norwegian State Housing Bank.[5]

Policies have to fit their local context: climate, culture, and built and natural environments. Policies to promote the construction of new low-cost housing in developing countries are different from those for the refurbishment of historic buildings in industrial nations. In many industrial countries, 80 percent of the existing building stock will be around for the next two decades and beyond, while developing countries are experiencing a new construction boom. The effectiveness of policy measures varies, depending on countries. Building codes, for example, are often successful in industrial countries, but less so in developing countries due to insufficient enforcement.[6]

Process. Process refers to the need for long-term thinking and translates into a life-cycle approach. Buildings are products of a complex demand and supply chain that can last for centuries, starting from the definition of a need and site selection. In parallel to different design stages, processes on the construction site are being managed, and purchasing decisions are made regarding structural and mechanical systems, construction materials, and products as well as on construction work. Services are procured for the longest phase in the life of the building—its use and operation—followed later by refurbishment and reuse, and finally deconstruction and recycling.

One fact is certain: the earlier in the process a decision is made, the bigger its impact. Researchers highlight the need for a devoted professional to coordinate the entire process and ensure that every choice helps implement the agreed targets. The sustainability coordinator's role could become a building permit requirement.[7]

The construction sector is infamous for shady deals. (See Box 10–1.) From the process perspective, the purpose of corruption is to avoid meeting agreed targets. It often starts when powerful interest groups lobby against the introduction of sustainability policies. The most efficient method for shifting construction onto a sustainable path in this regard would be the worldwide integration of the Project Anti-Corruption System (PACS) into project management. PACS uses a variety of measures that affect all project phases, all major participants, and a number of contractual levels. Information on corruption, anti-corruption tools, and examples of anti-corruption programs for gov-

ernments, funders, project owners, and companies are widely available.[8]

Performance. Performance underlines a holistic approach. A building does not become sustainable by adding up "green" materials and elements. Only the performance of the building during its entire lifetime matters. Hence policies are moving away from prescribing "foolproof" solutions, like telling the thickness of thermal insulation, for example, and instead toward asking for the minimum energy performance of the building.

There are strong expectations that hi-tech can meet the challenge. That is not going to happen. Innovations at the lo-tech end have much bigger impact, and their volumes are

---

Box 10–1. Examples of Corruption in Construction

False Invoicing through Supply of Inferior Materials. A concrete supplier is obliged to supply concrete to a particular specification. The supplier deliberately provides a cheaper product of inferior specification but invoices the contractor for the required specification.

Concealing Defects. A roofing subcontractor installs a waterproof roof membrane. The membrane is accidentally perforated during installation, which means that it could leak. It needs to be approved by the contractor's supervisor before it is covered over. The membrane should be rejected and replaced, but the subcontractor offers to pay the supervisor for certification that the subcontractor's defective membrane is water-tight. The supervisor accepts. The subcontractor submits the certificate to the contractor and obtains full payment for the defective membrane. Neither the subcontractor nor the supervisor tells the contractor that the membrane is defective.

Source: See endnote 8.

---

radically bigger. For example, the global trend toward thinner exterior walls has meant that building facades have no thermal mass and the need for air-conditioning has exploded. The solution is not more green air-conditioning equipment but buildings that perform better, with thicker walls, perhaps.

More than 600 rating systems to assess building performance are available worldwide. They cover everything from simple energy consumption evaluations to life-cycle analyses with an ecological focus and total quality assessments. There are also efforts to define a small set of core criteria. (See Box 10–2.) The systems have different tasks, depending on which questions they are supposed to answer. Some assess the predicted performance at the design stage, others the actual performance of the existing building. International investors demand different information than authorities measuring greenhouse gas (GHG) emissions. Property businesses use certification systems for branding.[9]

The increased export and import of the major assessment methods worldwide is also an export and import of their cultural underpinnings and has potentially adverse long-term consequences for promoting regionally specific practices. The selection of right performance levels and weighting criteria requires a good understanding of local conditions. If this is missing and the chosen criteria are too easy, the impact remains insignificant or even negative. A system with a number of different level indicators may be tempting for users who are more interested in easy credits than ambitious development.[10]

Sustainable Infrastructure. Buildings need sustainable infrastructures as their human-made environment. A wonderful "eco-house" is unsustainable if it gets electricity from a grid that distributes fossil-fuel-based energy. As policy tools, regional and urban plans secure a sustainable context for buildings if they integrate land use with mobility and key infrastructure.

Box 10–2. Searching for Core Indicators of the Sustainability of a Building

Sustainability advocates have made efforts to come up with 5–10 core indicators that are agreed on internationally, which could be complemented by local priorities. Universal criteria and benchmarks are urgently needed as instruments to support decisionmaking in new construction, procurement, and investment.

In 2009, six core indicators were presented to the Sustainable Building Alliance by a coalition of stakeholders:
- GHG emissions (Global Warming Potential)
  – CO_{2eq} (kilograms)
- Energy
  – Primary energy (kilowatt-hours)
- Water
  – Cubic meters
- Wastes
  – Hazardous (tons)
  – Nonhazardous (tons)
  – Inert (tons)
  – Nuclear (kilograms)
- Thermal comfort
  – Percent of occupied period where temperature exceeds a given value
- Indoor air quality
  – CO_2 in parts per million
  – Formaldehyde in micrograms per cubic meter

These six points cover the use of the two most basic resources and emissions, as well as indicators for the quality of the indoor environment. They are a good start, but even as core indicators of sustainable buildings, this is far from a complete list. The work continues.

Source: See endnote 9.

The goal is sustainable use of resources, prevention of urban sprawl, and minimization of mobility needs. Infrastructures are used efficiently when the water, wastewater, energy, communication, road, and mobility networks serve more people at shorter distances. Considering the huge share of CO_2 emissions that result from energy consumed in construction, heating, cooling, lighting, and appliances, recent policies promote the introduction of renewable energy sources: solar, wind, hydropower, geothermal, and biomass. Indeed, the shift toward more renewables is going to have a much bigger impact on GHG reduction than higher energy standards for new buildings and the refurbishment of existing ones (as imperative as those also are).[11]

Resource Use. Not all resources are equal. Some are renewable, others will run out at some point. Some are hazardous for human health, others pollute the air, atmosphere, soil, or water—the ecosystems needed for life. The production process of certain construction materials is extremely energy-intensive. Some construction methods are labor-intensive, others industrial.

More-sustainable use of resources is not only a technological question. Policies can create incentives also for savings or shifting to another resource. Saving often depends on behavior—learning to turn the water faucet off completely, for example. For behavioral change, tambourines are needed.

Shifting from nonrenewable to renewable fuels, even recycled waste, means simultaneously that air pollution is reduced and local resources and local labor can be used.

Resource efficiency can be supported by technology, like co-production of heat and electricity, without wasting primary energy. Incentives can support renovations where more-efficient appliances are installed, or cross-subsidies can lower the price of a bus trip.

Regarding energy, the Fourth Assessment Report of the Intergovernmental Panel on Climate Change was a wake-up call for the sector, as it reminded people that buildings offer the largest share of cost-effective opportunities for GHG mitigation among the sectors examined. The report highlighted a directive of the European Union as one of the most comprehensive pieces of regulation targeted at the improvement of energy efficiency in buildings. (See Box 10–3.)[12]

Conclusions

No single policy is going to change light green buildings into sustainable ones. Policy packages will have to combine sticks, carrots, and tambourines. Coordination, enforcement, and monitoring will be needed. Publications and websites will not suffice. It is also a good policy if a city decides to establish an information office for sustainable building or a state has an "energy consulting bus" driving around to meet people who need advice.

Measurable targets seem easy to define: net zero energy, zero carbon, and zero waste. Other targets need indicators, too: resource use, human health, access to basic services, decent work, and fair trade. There has to be agreement on the core criteria of sustainable building, and they will have to be applied to all decisions, including procurement and financing.

Mainstreaming sustainability starts with setting targets and doing preliminary designs, and it needs to be followed through until maintenance and performance monitoring. If corruption cuts any of the numerous links in the process, targets will be lost. The real difference comes from implementation of policies and having the patience to keep the targets in mind throughout a building's entire life cycle.

Box 10–3. European Union Directives on Energy Performance of Buildings

In December 2002 the European Parliament passed a Directive on the energy performance of buildings. It has four major components:

- It introduced a "common methodology for calculating the integrated energy performance of buildings," which may be differentiated at the regional level.
- Member states must apply the new methods to minimum energy performance standards to new buildings. The Directive also requires that a nonresidential building, when it is renovated, be brought to the level of efficiency of new buildings. This is important due to the slow turnover and renovation cycle of buildings and to the fact that major renovations of inefficient older buildings may occur several times before they are finally demolished. This is one of the few policies worldwide to target existing buildings.
- The Directive set up certification schemes for new and existing buildings (both residential and nonresidential), and in the case of public buildings it requires the public display of energy performance certificates. Informa-

tion from the certification process must be made available for new and existing commercial buildings and for dwellings when they are constructed, sold, or rented.

- Member states must establish "regular inspection and assessment of boilers and central air-conditioning systems" in buildings.

According to a more recent Energy Performance of Buildings Directive, as of 2021 all new buildings will have to consume nearly zero energy, and the energy consumed will have to originate to a large extent from renewable sources tapped by the building or in its vicinity. All buildings undergoing major renovation (25 percent of the surface) will need to improve their energy performance. The legislation required member states to list incentives, from technical assistance and subsidies to low-interest loans, for the transition to near-zero-energy buildings.

The European Union is also developing an Ecolabel and Green Public Procurement criteria for buildings.

Source: See endnote 12.

CHAPTER 11

Public Policies on More-Sustainable Consumption

Helio Mattar

A new middle class has emerged in Brazil thanks to the recent growth of the economy and the government's direct income distribution to the poorest Brazilians. The spread of the consumer culture is now a reality in this rapidly growing economy. More than 31 million consumers who joined the middle class—with an average family monthly income between $530 and $2,120—are now able to make choices in the products and services they buy. Just a few years ago, they were forced to buy only absolutely necessary products and services and at the lowest price possible. As a result of this growth in the middle class, this social group now accounts for 51 percent of total consumption in Brazil.[1]

A recent survey of women gives a good idea of the pattern of consumption in this new middle class—a pattern that closely follows that of richer Brazilians. After an initial stage of buying the products and services they really need—such as major home appliances, computers, and mobile phones—the women wanted to buy cosmetics and beauty products, invest in the aesthetic improvement of their teeth, and purchase a used car. And after the two initial stages of consumption preferences, the priorities of this new middle class

were to travel by plane, change the kitchen cabinets, put their children into private schools, dine out, and purchase broadband Internet service.[2]

It is clear that this pattern of consumption follows closely the unsustainable model of richer socioeconomic classes. And it is significantly shaped by the mass media, including television, which reaches practically 100 percent of the Brazilian population. The outcome of this onslaught of traditional advertising is no different from the overconsumption model that has spread around the world, also fueled heavily by advertising. In 2011, advertising expenditures hit $464 billion worldwide, 3.5 percent more than in 2010. Of these expenditures, not surprisingly, one third is spent in the United States—the leading consumer economy. Without significant changes to consumption patterns in Brazil and the entire globe, the planet will face increasing strains, and with it so will human society.[3]

Unsustainable Consumption

Worldwide, demand for consumer products has reached completely unsustainable levels. According to WWF's Living Planet

Helio Mattar is president of the Akatu Institute for Conscious Consumption in São Paulo, Brazil.

Report, the world demands 50 percent more renewable resources than the Earth can sustainably provide. In large part this stems from the massive material and energy demands that consumer societies put on natural resources.[4]

A report by the Sustainable Europe Research Institute, GLOBAL 2000, and Friends of the Earth Europe found that 60 billion tons of resources are now extracted each year—about 50 percent more than just 30 years ago. In 2000, someone living in North America used 88 kilograms of resources each and every day; in Europe, the figure was 43 kilograms of resources daily; and in Latin America, 34 kilograms each day.[5]

These materials are used not just for basics, like food, shelter, clothing, and transportation, but for the bevy of consumer products that have become a central part of so many cultures. In 2008 alone, people around the world bought 68 million vehicles, 85 million refrigerators, 297 million computers, and 1.2 billion mobile (cell) phones—numbers that will only continue to grow as more individuals enter the consumer class.[6]

In 2006, the 65 high-income countries where consumerism is most dominant accounted for 78 percent of consumption expenditures but just 16 percent of world population. As the remaining 84 percent of humanity tries to join the consumer economy, what type of actions could ensure that the new middle class pattern of consumption does not totally mirror that of today's top 16 percent? The push toward a more sustainable consumption system has three elements: technology change on the part of companies, behavior change on the part of consumers, and public policies that provide incentives for both these changes. The good news is that increasing amounts of research find that well-being is rarely connected to consumption. (See Box 11–1.)[7]

Putting Pressure on Corporations

In August 2011 a Nielsen Internet survey found that 83 percent of respondents around the world said that it is important for companies to implement programs that improve the environment. Unfortunately, only 22 percent of those surveyed also said they would pay more for environmentally and socially sustainable products. Still, recent work by the World Economic Forum on sustainable consumption shows that corporations are already well aware of the need to change the consumption model.[8]

Civil society is playing a key role in getting corporations to understand the need for change. Nongovernmental organizations and pressure groups have been challenging everyday social norms—whether through local efforts like purchasing cooperatives, efforts to share resources through perhaps a tool library, or campaigns to put direct pressure on corporations. The Rainforest Action Network, for example, mobilized thousands of activists to pressure Home Depot into using more-sustainable forest products. And Greenpeace got people all over the world to rage against Nestlé for its use of palm oil from companies that, according to Greenpeace, were destroying Indonesian rainforests, threatening the livelihoods of local people, and pushing orangutans toward extinction. As a result, Nestlé announced a commitment to identify and exclude "companies from its supply chain that own or manage 'high risk plantations or farms linked to deforestation.'"[9]

Digital networks all over the world are also raising the awareness of consumers about the social and environmental impact of consumption. A GlobeScan 28-nation poll in July 2011 revealed that "regular users of Facebook, Twitter and other online social media expect higher levels of corporate responsibility from companies, and are more likely to act on their values as ethical consumers." The

Box 11–1. Consumption, Communities, and Well-being

Although people have long known that "money can't buy happiness," many of them still literally buy into this idea. Yet scientific research shows that, over the long term, the purchase of more "stuff" does not make individuals happier or healthier.

Of course, there are some real correlations between the ability to meet basic needs and subsequent well-being, but beyond that a great deal of a person's life quality stems from health, social relationships, and meaningful work. Moreover, high levels of consumption are increasingly undermining health and community ties as more people work longer hours and spend more time in cars and in front of television and computer screens.

Considering the deep disruptions that human consumption patterns have caused to Earth's environment, the goal should be to shatter the myth that stuff brings happiness and instead actively pursue policies that get the most human well-being out of every unit of natural resource. The Happy Planet Index examines how this could be done, comparing well-being levels of different countries with their ecological impacts. The United States, China, and India all experienced a drop in their happiness scores over the last 15 years. Costa Rica, on the other hand, has the happiest people per hectare of resource used. Costa Rican economist Mariano Rojas attributes this to strong community networks, facilitated by a balanced work-life culture.

Rebuilding strong relationships with families, friends, neighbors, and local communities will be part of a key strategy for increasing the well-being of both the environment and human societies. This not only improves quality of life, it also helps substitute social capital for financial and natural capital. One study shows that co-housing and ecovillage residents who have strong community ties report life satisfaction levels equal to the residents of Burlington, Vermont, who share similar demographics but earn twice the annual income.

Replacing things with relationships and shared community resources provides opportunities to reduce consumption. Tool libraries, toy libraries, and shared community spaces encourage social networks and enable people to comfortably reduce their home sizes and the amount of things they own. In Columbus, Ohio, the town's tool library has over 4,000 members; in nine months it loaned 3,043 tools to 933 individuals and 1,946 tools to 98 nonprofit groups, saving residents and local organizations hundreds of dollars on average. In New Zealand, 217 toy libraries provide a variety of educational toys for over 23,000 children.

Governments around the world are now starting to incorporate well-being measures into policy as well. The U.K. government has added subjective well-being measures to its set of sustainable development indicators, while the government of Wales has included the ecological footprint in its top five indicators of sustainability. Currently the European Union is considering doing the same. Governments globally need to follow suit, incorporating policies that will maximize human well-being while minimizing ecological impacts.

Skill sharing, relationship building, and community participation are the seeds of trust, community, and a truly sustainable well-being.

—*Dagny Tucker*
Universitat Jaume I Castellon de la Plana, Spain
Source: See endnote 7.

poll found that 31 percent of regular social media users said they had rewarded a socially responsible company, compared with 24 percent of those who do not use such media regularly. And 23 percent of social media users said they had punished a socially irresponsible company by criticizing them or boycotting their products, compared with 17 percent of non-users. "This group is also more likely to say they regularly choose to pay extra for environmentally friendly or ethical products and services, only buy from responsible companies, and that they think socially and environmentally friendly products are of higher quality," GlobeScan reported.[10]

The poll also found that social media users are "more likely to possess opinion-leader characteristics, such as being in leadership positions at their workplace or community, support [an] NGO, and frequently discuss business and politics." It confirmed that "a new generation of consumers is turning to less traditional, unofficial sources of information on CSR [corporate social responsibility], such as social networks like Facebook or Twitter, while company websites are being left behind by consumers looking for CSR information." GlobeScan senior vice-president Chris Coulter concluded that companies "can no longer afford to ignore social media as a channel for communicating their messages around corporate responsibility. Users are more switched-on to ethical business, more empowered, and more influential—and as people look beyond traditional sources of information on corporate responsibility, their attitudes are shaping those of others."[11]

Transparency brought about by the digital networks has changed the visibility of all corporate actions and omissions. A Brazilian company, Arezzo, recently launched some products that used the fur of foxes and rabbits, for example. Consumers showed their outrage at this in the "twitterverse." The company argued that the foxes and rabbits were raised in compliance with respected international certification standards. Not good enough, said those using the digital networks. They expressed their repulsion at the idea of killing animals just to use their skins, and Arezzo took the line of products off the market.[12]

Providing Incentives, Pushing for Change

The pursuit of sustainability and sustainable consumption will require a concerted effort on the part of all—from governments and producers to civil society and consumers themselves. In the face of significant advertising budgets, mass media influence, an ingrained and growing consumer culture, and the sheer environmental demand, good policy will be not only appreciated but crucial to the environment and the very future of society.

Changing consumption behavior requires changing a very important part of the culture of any society. For that to happen, it is necessary to change the socially valued patterns of behavior so that sustainable consumption gains social recognition and validation until it becomes the new norm. Given their reach and power, public policies must be used to help influence consumers' behaviors so as to increase the speed of change in society's perception of the desirability of more-sustainable consumption.

Perhaps the most important public policy change needed is to reduce taxes on more-sustainable products and services or increase taxes on the lesser ones. Price makes a difference to consumers. A survey in 10 industrial countries and 7 emerging ones found that when consumers were asked which aspects of a product are important in buying decisions, 80 percent point to quality, 72 percent mention price, and 45 percent cite social and ethical aspects of the companies. Examples of this public policy can already be found. In Sweden, for instance, "green" cars are exempt from a vehicle tax for five years, and that tax for all cars is adjusted in light of the amount of carbon dioxide a particular vehicle emits. A related important public policy would be the internalization of externalized costs, as Australia did recently with a tax on carbon emissions.[13]

Education for sustainability and for sustainable consumption in public schools is also important. Teaching children, starting at a very early age, about the positive and negative impacts of consumption on society and the environment is a very effective way to establish more-sustainable patterns of consumption early on in people's lives. And children can in turn influence their parents' behavior. The Akatu Institute developed a series of 10 animation videos on the themes of sustainability and "conscious consumption" along with training materials for teachers; this has been used in more than 1,500 schools, where teachers learned about the theme and took it to their junior high school students.[14]

Media literacy is related to this topic of education. It is important to create critical analyses of the commercial messages that consumers are routinely exposed to. This highlights the point that the transition to sustainable cultural practices can only happen if people learn how to critically engage with the media.[15]

Given the enormous visibility and purchasing power of governments, it is vital that they lead by example. One good example of this is the city of São Paulo, where outdoor advertisement was banned, improving considerably the quality of the urban environment and at the same time significantly reducing people's exposure to commercial messages. As a result, 15,000 billboards were removed. While some critics warned that this would cause irreparable damage to the city's economy, there were no adverse consequences, and today 70 percent of city residents view the ban as having improved the city.[16]

Considering how much governments themselves consume, another important tool is the introduction of sustainability criteria in public procurement processes. Often this starts at a local level. In San Francisco, the Precautionary Purchase Ordinance requires the city to take environmental and health concerns into consideration when making purchases. The incentive for manufacturers to meet procurement requirements increases significantly when this type of policy is in place. In addition, the prices of more-sustainable products and services can come down due to the economies of scale associated with large government purchases. Ensuring that material flows are circular—and that post-consumer waste becomes the resource for the next generation of products—is also an important role for governments to play. (See Box 11–2.)[17]

Companies could receive incentives to use their advertising and product packaging to educate consumers toward more-sustainable consumption. In Brazil, people have been talking about banning advertising directed at children for 10 years, but there is still no law on this. Yet the fact that it is being discussed caused the food industry to decide on its own to only advertise food and drinks that were deemed nutritional according to criteria dictated by scientific evidence. In June 2011, the Brazilian Advertising Self-Regulation Council (known as Conar) imposed quite strict rules on companies claiming to be sustainable or to have sustainable products. In addition to prohibiting advertising containing any incentives for pollution and waste, the new rules say that any environment-related claims of companies need to follow four principles: verifiable claims, precision of claims, pertinence of claims to specific industrial processes and products, and relevance of the environmental benefit considering the full life cycle of the product.[18]

Governments have power to direct media efforts toward education for sustainable consumption. These could have a rapid and sustained effect on consumers' behavior. This was the case with the educational campaign of the Ministry of Environment in Brazil to reduce the use of plastic bags. It started in June 2009 in an initial partnership with Walmart and later with Carrefour through which 19 radio spots and three films for television and movie theaters showed in a creative way the

Box 11–2. Japanese Efforts to Build a Sound Material-Cycle Society

In 1994 Japan's Basic Environment Plan recognized that socioeconomic activities characterized by "mass production, mass consumption, and mass disposal" were a common driving force of various environmental problems. The transition to a "sound material-cycle society" (SMCS) became a top environmental policy priority.

Early efforts focused on recycling, as this was an immediate way to start shifting to a circular society and reduce total solid waste flows, which due to a shortage of landfill space was an urgent issue. Waste sorting by consumers for recycling is a visible and easy-to-understand action, and its promotion helped reduce the actual amount of waste as well as publicize the need to improve consumers' traditional wasteful consumption patterns.

Several recycling laws for specific product categories have been enacted and enforced. They cover containers and packaging, home electric appliances, end-of-life vehicles, food waste, and construction waste. Numerical targets such as recycling rates for these product categories were set, and their progress has been reviewed regularly. Today, 78 percent of PET bottles and 77 percent of wastepaper are collected for recycling in Japan—up from 2 percent and 53 percent respectively in 1995.

Over time, SMCS policies focused less on recycling and "downstream" treatment of waste and more on linking waste to resource issues upstream. Waste and resource issues are often discussed separately, managed by different authorities, handled by different industries, and studied by different schools. Both upstream and downstream problems can be solved in a win-win manner if integrated approaches are taken to manage materials throughout their life cycles.

As material resources such as metal ores are becoming scarcer, there is an increasing rationale for upstream industries, such as smelters, to seek a secondary supply of resources from recycling activities. One example of this is DOWA, a company that applies its advanced technologies, originally developed in mining and refinement operations to extract precious metals such as gold and silver, to recycling of as many as 17 different metallic elements from end-of-life products.

Japanese environmental policy had never explicitly addressed the need to save natural resources before the enactment of the Basic Act Establishing a Sound Material Cycle Society in 2000. The act indicates that SMCS means a society in which the consumption of natural resources will be conserved and the environmental load will be reduced to the greatest extent possible. As the Japanese economy depends heavily on imported resources, negative environmental impacts caused by extraction and harvesting of natural resources in foreign countries have been mostly hidden. The SMCS, at least conceptually, incorporates such indirect impacts of Japanese activities. For example, the second fundamental plan for SMCS adopted an indicator to monitor changes of hidden flows associated with the import of metallic resources.

While there has been progress in implementing SMCS policies, the 2011 tsunami and resulting nuclear catastrophe disrupted the circular loop for many waste streams—with debris, municipal waste, and sewage sludge being polluted by radioactive fallout. This, and the cleanup of the area affected by fallout, will bring significant new challenges to Japan's efforts to make a sound material-cycle society.

—Yuichi Moriguchi
University of Tokyo
Source: See endnote 17.

environmental impacts of plastic bags. The initial goal of the campaign was surpassed: after 10 months, an estimated 5 billion bags were saved—that is, not distributed. Another interesting form of government engagement would be to tax advertising and using some of that revenue to sponsor counter advertising to market a sustainable lifestyle. Alternatively, governments can help lower overall consumption pressures by reducing advertising altogether—either on television, as Sweden has done for children's programming, or other forms, like outdoor advertising, as São Paulo has done.[19]

Governments can also push companies to continuously improve their products—making every generation of products more sustainable. In Japan, the government's Top Runner program encourages continued innovation by frequently testing products for efficiency. The most efficient become the baseline standard for the next generation of product, thus continually pushing companies to produce more-efficient products. However, this standard currently only applies to products intended for the domestic Japanese market, so it does not mean the rest of the world is getting more efficient Japanese products.[20]

Companies that operate and develop products in a way that makes an important contribution to sustainability can be publically recognized by governments. One contribution companies can make is allowing employees to work at home and have flexible and fewer working hours. This can reduce overall consumption by eliminating some trips to and from work and at the same time let people enjoy more of their intangible assets, such as relations with friends and family.

An excellent example is the work-life balance awards administered each year through New Zealand's Equal Employment Opportunities Trust, which recognize some of the best practices in work and life. The Prime Minister gives the awards to companies at a

gala dinner, and the practices and policies of all entrants are covered in New Zealand's Best Employers. One of the award categories, "Walk the Talk," recognizes senior managers "who act as champions and enable employees to improve their work-life balance." To qualify for the award, organizations must "provide evidence that the work-life balance policies are benefiting employees" and, in large organizations, "the initiatives must be integrated into the organizational strategy, culture, practice, senior management accountability and measures of success."[21]

National indicators for well-being would show consumers the value of a sustainable lifestyle, which would help change consumption behavior in the direction of a more balanced life–work relationship and toward sufficiency in consumption. (See also Chapter 6.) An excellent example is the Gross National Happiness index and goals developed by the government of Bhutan. These served as inspiration for the proposal for a Commission on the Measurement of Economic Performance and Social Progress, established in early 2008 by President Nicolas Sarkozy of France. The Commission has proposed new indicators to measure subjective aspects of social progress such as freedom, security, and contentment as well as objective features including economic and ecological resources.[22]

The Commission was set up because of a clear perception that the official statistics on economic growth do not reflect the way people perceive the conditions of their lives. Current indicators twist the political debate and any consequent actions away from the real needs of human society. The Secretary-General of the Organisation for Economic Co-operation and Development, Angel Gurría, welcomed the Commission's recommendations: "Economic resources are not all that matter in people's lives. We need better measures of people's expectations and levels of satisfaction, of how they spend their time, of their relations with other people in their community. We need to focus on stocks as much as on flows, and we need to broaden the range of assets that we consider important to sustain our well-being." Another effort to focus on stocks of resources and how they are consumed—this time at the global level—is the move to establish Millennium Consumption Goals. (See Box 11–3.)[23]

Providing information on the sustainability of products all along their supply chain is a very expensive endeavor. Governments could help make such information more widely available so that consumers could make better product choices. This would also serve as an educational tool to raise consumer awareness about the impacts of their choices. Many governments already participate in certification processes to help consumers identify such products as organic produce. The private sector has begun to develop other tools in various formats, and there is a great deal of room for further government support.

The "Good Guide," an online resource that can be used on smartphones, rates products based on three categories: health, environment, and society. Users can quickly find data on thousands of products in order to make better choices—even if they are still not perfect ones. Other tools to help consumers navigate the overwhelming number of products available would be of great assistance to the 83 percent of Nielsen survey respondents who say it is important for companies to take steps to improve the environment. With more information, these individuals can find out which companies best meet this goal.[24]

Only a society that is aware and mobilized toward the need for sustainability will have

---

Box 11–3. Setting Global Goals

There has recently been an effort at the global level to foster a dialogue about overconsumption and the need to reduce this worldwide. A small group of civil society organizations is working to establish Millennium Consumption Goals that help create targets for reducing overall consumption around the world, particularly in consumer populations. These goals are offered as a complement to the Millennium Development Goals—a series of eight goals to reduce global poverty—by reducing consumption to free up ecological space in order to pursue poverty alleviation measures.

While it is early in the development of this effort, leaders of this initiative have presented the idea and are working to include it on the Rio 2012 conference agenda. The group also continues to develop specific concrete goals, considering proposed reductions in obesity rates, motorized transport, military spending, total energy use, length of the work week, and income inequity.

—Erik Assadourian
Source: See endnote 23.

---

the strength and persistence to bring sufficient pressure on governments to enact and adopt public policies that reflect sustainability as a real priority. So education for conscious consumption is absolutely necessary in order to break the vicious cycle of governments accepting myopic short-term lobbying for today's unsustainable production and consumption patterns.

CHAPTER 12

Mobilizing the Business Community in Brazil and Beyond

Jorge Abrahão, Paulo Itacarambi, and Henrique Lian

Twenty years after the first Rio Conference, the world has changed dramatically. Global population has grown 28 percent, the global economy has expanded 75 percent, and Earth's systems have become more strained than ever before. The global economic downturn that started in 2008 revealed how strongly short-term economic imperatives outweigh political decisions, even as the traditional development model confronts the need for a dramatic change of course based on sustainable development principles.[1]

The 1992 Rio conference did bring about significant successes, producing a robust set of agreements among nations as expressed under the 27 Principles of the Rio Declaration, *Agenda 21*, the Declaration of Forest Principles, and the conventions on biological diversity, climate change, and desertification. It also opened the way to later agreements, such as the Millennium Declaration and the Millennium Development Goals, the Johannesburg Plan of Implementation, the Latin American and Caribbean Initiative for Sustainable Development, the Monterrey Consensus of the International Conference on Financing for Development, and the Bali Strategic Plan for Technology Support and Capacity-building. *Agenda 21* has proved to be a strong driver at the regional and local levels, contributing to building strategies and policies for more-sustainable communities.[2]

But for all that the first Rio summit accomplished, it has not helped turn humanity away from the unsustainable path it is on. Indeed, given the shift in economic realities, addressing sustainable development independent of economic priorities will be a recipe for failure. Hence, it is no surprise that one of the themes of Rio 2012 is "a green economy in the context of sustainable development and poverty eradication." Yet what exactly does that mean?[3]

A new model of production and consumption must embrace planetary limits, the need for reducing inequalities of income and opportunity, preservation of the rights of future generations, ethical principles, and a whole new paradigm of development that is not based merely on economic growth. Despite the "extraordinary run since the start of the industrial revolution two centuries ago, lifting billions of ordinary people out of abject poverty," according to its advocates, the current eco-

Jorge Abrahão is president, **Paulo Itacarambi** is vice president, and **Henrique Lian** is head of institutional affairs at the Ethos Institute in Brazil.

nomic model stands revealed as socially not inclusive, as environmentally predatory, and as placing private interests above public ones. In short, this model is unable to address the needs of a world with 7 billion people, climate change, and alarming levels of poverty.[4]

Green, Inclusive, and Responsible

Given these challenges, it is necessary to develop a road map toward a green, inclusive, and responsible economy:

- A green economy seeks to reconcile society's production processes and natural processes—promoting conservation, restoration, and the sustainable use of ecosystems and treating the services they offer as assets of public interest.
- An inclusive economy seeks to meet the needs and the rights of all human beings, promoting a better balance among financial, human, social, and natural forms of capital, a more equitable distribution of wealth and income-generating opportunities, fair access to public goods and services, and decent life conditions for everyone.
- A responsible economy seeks to strengthen a set of humanistic and universal principles and values that sustain the democratic functioning of societies and markets through the development of ethical and integrity values, promoting a culture of transparency and mechanisms to fight corruption.

Understanding that a new economy must establish a new relationship between society and nature, respect limits to growth, and embrace a permanent innovation process that is oriented toward sustainability, it is clear that new patterns of social and industrial metabolism—and indeed an ethical approach—are urgently required. Ricardo Abramovay of the University of São Paulo argues that "this challenge must be met neither by the State monopoly over business decisions nor through the abolition of markets. On the contrary, it must

be addressed in the context of a decentralized economy in which markets play a decisive role although not an exclusive one."[5]

Although defining key elements of a green, inclusive, and responsible economy may be a straightforward exercise, creating an economy that meets these ideals is a much greater challenge. A number of key steps to internalize existing multilateral commitments in local economies could help bring the world closer to this ideal.

Adopt a New National Accounting Standard. The United Nations needs to develop a new accounting standard that can be adopted by all nations. It should redefine the concept of prosperity, considering not only gross domestic product (GDP) measurements but also the costs of natural assets and services implied in the production of goods and services, the social impacts of the prevailing growth model, and access to adequate sanitation, health, education, consumption, mobility, culture, and well-being. The new national standard should measure natural, social, human, and financial capital along the lines developed by the Commission on Measurement of Economic Performance and Social Progress, which was chaired by Joseph Stiglitz. The Commission recommended improvement of numerical metrics on health, education, personal activities, and environmental conditions and urged the development of reliable tools and indicators.[6]

Move toward Carbon Pricing. Carbon pricing is indispensable to emissions control. All nations should adopt policies to facilitate the creation of national carbon markets. To ensure that national emission-reduction goals are achieved, it is essential that the characteristics of local markets and economic forces are studied carefully for the design of carbon pricing policies. Recent encouraging policy initiatives include a tax on carbon approved by the Australian Congress in November 2011, the European Union's emission-trading system, and

China's move toward an experimental domestic carbon market.[7]

Pay for Ecosystem Services. Appropriate pricing of natural resources and environmental services is critical in order to change individuals' perceptions and the way markets function. The goal is to close the production loop and fully acknowledge the shared benefits derived from biodiversity and traditional forms of knowledge. Some enlightening studies have attempted to estimate the value of ecosystem services. The first important survey was published in 1997 in *Nature*, and the authors found that the world's ecosystem services were valued at $33 trillion a year—more than the entire global economy at the time. (See Chapter 16.)[8]

The anole Anolis transversalis *photographed in Yasuni National Park, Ecuador*

Geoff Gallice

In addition to pricing carbon, a variety of efforts to price ecosystem services have been made—from creating common assets trusts to paying farmers to plant trees and direct payments to individuals for preserving an ecosystem and its services intact. In 2011, Ecuador, for example, said it would not exploit the 900 million barrels of oil in the ground under Yasuni National Park—a tropical rainforest that may be one of the most biodiverse places left on the planet—in exchange for $3.6 billion in aid for community development and renewable energy projects.[9]

Establish Minimum Operating Standards. Whether they operate domestically or internationally, companies should be required to adhere to a set of standards with regard to decent work, inclusion of minorities, and socio-environmental practices compatible with sustainable development and closed-loop production. Instead of promoting a global race to the bottom, multinational corporations should be encouraged to operate everywhere according to their own best national standards, in an effort to improve local standards. Global

reporting standards are essential, as is the requirement for annual public disclosure of sustainability activities. (See also Chapter 7.)[10]

Promote Sustainable Production and Consumption. Sustainable government procurement policies, R&D programs, and tax regimes can encourage forms of production that put less pressure on natural resources, entail low emissions, and allow decent work conditions. A study by ICLEI–Local Governments for Sustainability found that governments can reduce environmental impacts significantly by shifting public purchasing choices. For example, Europe could get 18 percent of the way to its Kyoto commitments just by the public sector committing to purchase only renewable energy. With Brazilian public procurement representing about 10 percent of that country's GDP, a shift in these policies could be a strong driver pushing the internal market toward more-sustainable products.[11]

A Brazilian adaptation of ICLEI's original study was developed in a partnership between the Local Governments for Sustainability Network, ICLEI's Office for Latin America and the Caribbean, and the Center for Sustainability Studies at the School of Business Administration of São Paulo's Getulio Vargas Foundation. Its second edition was also supported by the governments of the states of São Paulo and Minas Gerais, in addition to the Municipality of São Paulo, which seems to be a promising start.[12]

Sustainable production patterns need to be paired with behavioral changes on the part of consumers. For example, the EthicMark Award for Advertising that Uplifts the Human Spirit and Society was created in 2009 to "help advertisers and corporations accept their huge responsibilities in our democracy for educating the public on their choices and in directing the content of programming in a positive direction." One of the 2011 winners, the Nike Foundation, was recognized for its efforts to help empower girls to be young "champions" breaking the cycle of childhood poverty around the globe. The World Business Academy, one of the organizations behind the EthicMark award, encourages companies to refrain from "neuro-marketing"—a manipulative new form of marketing research that uses brain-scanning technology to better evoke certain emotional responses. Hundreds of companies have so far signed the pledge.[13]

Invest in a New Education Model. A new education model is needed that promotes awareness of sociocultural heritage, develops a culture that values the environment, and promotes people's sense of responsibility as citizens, voters, parents, consumers, investors, and entrepreneurs. While much is still to be done on this front, there are some promising models to make education more relevant for life in a green, inclusive, and responsible economy. In countries from Argentina and Australia to South Korea and Sweden, media literacy is being taught to students and youth. Australia, Canada, and New Zealand in particular have made great strides, incorporating media literacy into their core curricula and building collaboration between the media industry, educators, and regulators. UNESCO has

played an active role in media literacy, training educators in developing countries to ensure smart engagement with the media worldwide. To promote environmental awareness, UNESCO has sponsored university chairs in sustainable development. Sustainability education has been integrated into curricula at 45 universities in 27 countries.[14]

Promote Sustainable Cities. Policies and regulatory tools can help bring about suitable infrastructure investments, improvements in sanitation, a cleanup of water resources, sustainable transportation systems, and diversified, renewable energy generation systems. One example of this is the Sustainable Cities program launched in 2011 by three Brazilian organizations—Our São Paulo Network, Sustainable Cities Network, and Ethos Institute—aimed to raise public awareness and nudge Brazilian cities to develop in an economically, socially, and environmentally sustainable way. The initiative incorporates social, environmental, economic, political, and cultural dimensions, and it provides a set of indicators for governance, equity, and sustainability. The program also aims at strengthening transparency and social control.[15]

Establish an International Fund. International funds will be needed to support the national sustainability plans of many countries. Resources could be generated through a variety of means, such as an assessment based on the proportional contributing capacity of U.N. member states, allocation of 0.7–1 percent of industrial countries' GDP, an auction of the rights to use maritime and air space, or a tax of 0.05 percent on speculative international financial transactions (known as the Tobin Tax). The last one was first proposed by Nobel Laureate economist James Tobin in 1972 and has received a growing number of endorsements, including from philanthropists Bill Gates and George Soros, former U.S. vice president Al Gore, Pope Benedict XVI, and German chancellor Angela Merkel. In the words of Chancellor Merkel, "a financial transaction tax would be the right signal to show that we have understood that financial markets have to contribute their share to the recovery of economies." Financial resources would then be allocated in accordance with countries' voluntary sustainability commitments, subject to independent oversight. Countries with stronger goals on carbon emissions, biodiversity, poverty, and inequality, for instance, would be granted more funding.[16]

Implementing the Vision

While the policy elements just described represent a significant redirection of the global economy, they are essential steps considering the dire state of the world today. The corporate social responsibility (CSR) movement can help overcome resistance from the market and society at large. In Brazil, the Ethos Institute has been working with companies to influence the corporate environment. After 10 years of intensive efforts, the limits of the movement at the market, society, and values levels have become clearer.

First and foremost, the market has not developed effective mechanisms for rewarding or punishing companies on the basis of CSR criteria. The space for differentiating companies in the eyes of the market is still very marginal, as demonstrated by the limited investor response to new tools like the Bovespa Sustainability Index. While the private sector typically prefers self-regulation, it remains to be seen whether such an approach alone is sufficient.[17]

Second, the culture of sustainability is not yet mature enough to compel sufficient changes in corporate behavior deeply. Lack of information, knowledge, and interest—or its superficiality—put the media, companies, business schools, and citizens all in a position of comfortable passivity.

Last but not least, ethical and fundamental human values are peripheral to the corporate decisionmaking process. Efficiency, low costs, high profits, and large scales of production still matter much more than sustainability values essential to the welfare of present and future generations.

Transcending these barriers requires that civil society organizations continue to engage the business community yet also work toward better regulations. One activity will influence and, it is hoped, reinforce the other.

Recognizing this, since 1998 the Ethos Institute has played a convening role in bringing together civil society and business actors to wrestle with major environmental and social issues—from climate change and solid waste to corruption and human rights. (See Box 12–1.) Working groups established by the Institute now involve more than 130 companies.[18]

The Climate Forum, for example, emerged in the lead-up to the December 2009 Conference of the Parties to the climate treaty. The goal was to signal to the Brazilian government that some big companies were willing to voluntarily reduce their greenhouse gas (GHG) emissions and to encourage the country to take a leading role at that meeting in Copenhagen. The group realized that reducing emissions is both an ethical proposition and an aid to Brazil's competitiveness. This helped influence the approval of the National Policy on Climate Change, which transformed the country's international voluntary commitment into national policy and subsequently helped ensure passage of a bill aimed at establishing sectoral plans for GHG emission reductions.[19]

The Solid Waste Business Forum aims to contribute to implementation of the National Solid Waste Act in São Paulo. The group has worked to expand general knowledge about the national policy, taking into account social and economic impacts, in order to guide corporate actions. It has established commitments that help implement national legislation. It also works to ensure proper handling of recyclable materials and better integration of col-

Box 12–1. The Roots of the Ethos Institute

Fourteen years ago, a group of entrepreneurs in Brazil—encouraged by a few visionary top company executives—began to articulate a movement aimed at sharing a vision in which business is a strong force for positive social change. This led to the launch of the Ethos Institute of Business and Corporate Social Responsibility, whose goal is to "mobilize, encourage and help companies manage their business in a socially responsible way, making them partners in building a sustainable and fair society."

Since its inception the Institute has encouraged and supported companies to change their management standards—incorporating social, environmental, and ethical concerns in their decisionmaking processes. The Ethos Indicators, now in their third iteration, have provided a road map for the implementation and assessment of social responsibility and sustainability principles in the companies' management. The Institute has also promoted the adoption of interna-

tional tools and standards such the Global Reporting Initiative and the Global Compact Principles. Several working groups are pushing for better public policies. Moreover, the Institute has mobilized other actors—local and national governments, consumers, civil society organizations, trade unions, the scientific community, and the media—to push companies and reward them for responsible practices, such as through the Climate Forum.

Recognition of the limits of the movement—reflected in the difficulty in changing values in markets, companies, and society—led to a deep reflection from which another strategy has emerged. This new step aims to formulate, through a collaboration of civil society organizations, a national project of sustainable development to be run by governments and a global movement for sustainability. The latter is benefiting from the increasing interaction among people around the world in relation to Rio+20.

Source: See endnote 18.

lection into companies' value chains.

The specific objectives of the Business Pact for Integrity and Against Corruption are dissemination of information on applicable legislation and encouragement of transparent and lawful contributions to political campaigns. Recently this working group has pushed for a set of laws to promote a culture of integrity—through, for example, bills on lobby regulation, civil liability of companies that practice acts against the public administration, and access to public information—a new regulatory framework for public procurement, and the implementation of a positive list of companies (Clean Companies Data) developed in partnership with the Federal Investigations Bureau.[20]

The Business and Human Rights group is supporting efforts to promote gender and

race equality in the workplace, eradicate slave labor in the value chains of companies, include people with disabilities in the labor market, ensure the rights of children and youth, strengthen social dialogue, and create decent work conditions. It also focuses on the promotion of a stronger social dialogue guided by the guarantee of human rights and creating decent work.[21]

In its most recent initiative, the Ethos Institute has brought together representatives from 35 organizations—including business associations, trade unions, government agencies such as the National Bank of Development, universities, and civil society groups like Greenpeace and WWF—for a Transition Committee to debate each of the pillars and strategies for implementation of an inclusive, green, and responsible economy. Discussions of the Transition Committee at the annual Ethos Conference in August 2011 received extensive media coverage.[22]

The issues debated at the conference—which brought together hundreds of companies and many representatives from the Brazilian federal government (including ministers and secretaries of state)—were wide-ranging and included governance for a new economy, new patterns of production and consumption for sustainability, innovation for sustainability, impacts of the new forest legislation, human rights, financing for the new economy, energy, biodiversity, solid waste, climate change and its impacts in the new economy, infrastructure for a new economy, extreme poverty eradication, decent work and green jobs, education for sustainability, water management, sustainable cities, integrity and transparency, and Rio+20. The conference underscored the need for new public policies able to foster sustainable development in areas such as energy, water, transportation, biodiversity, and cities.[23]

Where the political environment inhibits regulations that could move societies more quickly toward a green economy and increased corporate responsibility, civil society should continue to push collaboration forward. Yet more regulation, and better regulation, is needed to lead private investments as well as to guide a more mature social dialogue, allowing the private sector and civil society organizations alike to take part in the regulatory process and thus lending it legitimacy, realism, and enforceability. Voluntary commitments are essential first steps. But over time they need to become legally binding. Exemplary behavior by some companies helps to raise industry standards and encourages governments to regulate.

Under the banner Global Union for Sustainability, the Ethos Institute is working to mobilize various segments of society for a global sustainability movement, promoting both dialogue and action. Leaders of business, labor, academia, women, youth, indigenous and traditional communities, farmers, local authorities, and nongovernmental organizations are being asked both to undertake voluntary commitments and to press for strong regulations in their countries and in terms of global governance.[24]

First discussed during a working meeting in Rio de Janeiro in October 2010, with the participation of 100 Brazilian and international leaders, the Global Union initiative has recently been introduced to important international forums. An International Steering Committee was going to be set up in January 2012 and a first set of commitments will be launched at Rio+20.

Ensuring That Rio 2012 Fosters a Green Economy

The world's major sustainability conferences—Stockholm in 1972, Rio in 1992, Johannesburg in 2002—have taken place in very different economic circumstances. The Stockholm Conference took place as the Bretton Woods institutions lost effectiveness, soon to be followed by two oil crises (1973 and 1979) that shook the global economy. Rio 1992 was staged against the backdrop of economic deregulation, the fraying of national frontiers, and reduction of social protection networks—all of which contradicted the sustainable development effort.[25]

The agreements forged in Rio in 1992 were of a fundamental nature, but they were not implemented as they contradicted the logic of growing globalization. The Johannesburg Summit was convened as the movement of capital flows through the global economy expanded to new heights, when the world's capital was allocated predominantly in service of its own reproduction. At each of these historic moments, there was a disparity between the summits' proposals and deliberations and the daily decisions actually taken by governments and businesses. Conventional economic logic took precedence over the political agreements.

Rio+20 faces the challenge and the opportunity presented by a conventional development model that has exhausted itself. Yet short-term economic imperatives once more are being given priority over political decisions for the long-term well-being of humanity.

This conference can be successful if it accomplishes four critical tasks. First, it should reaffirm the commitment of nations to sustainable development and previous multilateral agreements. Second, the participants should outline a new governance model for sustainable development, with strong participation of all major societal groups, that translates into a new Council for Sustainable Development. The Council should be formed by the institutions within the U.N. system that are responsible for its different dimensions, including

the financial and justice ones, and should be comparable to the Security Council.

Third, the conference should encourage countries to formulate national plans of sustainable development adapted to different local realities. A minimum agenda for these should include goals to reduce ecological footprints, to eradicate poverty and reduce societal inequalities, and to implement a system of integrity and transparency in order to translate commitments into the real economy and real politics. Fourth, Rio+20 should outline new mechanisms of financial support for implementation of the national plans.

Success also depends on a greater involvement of civil society. In the event that governments do not move forward, society at large cannot be paralyzed as well. Many private and social actors are looking for ways to participate in this conference. Rio+20 needs to capitalize on these efforts. By empowering nongovernmental actors, the conference itself will be empowered.

CHAPTER 13

Growing a Sustainable Future

Monique Mikhail

There is a growing global consensus that the world's food and agriculture system is broken. The good news is that solutions exist and are beginning to take root. Yet it will take a concerted effort by a variety of actors at the local, national, and even global level to bring about a sea change in the way we nourish ourselves in the context of increasing planetary resource constraints.

There are a host of "different worlds" in agriculture, and it is not just "large" or "small" that matters. Even within smallholder agriculture, a wide variety of physical, social, and economic conditions require different targeted solutions. There are also different worlds in terms of geography and the role of smallholder agriculture within the wider political economy. Generalized prescriptions for sustainable agriculture do not work because the starting points are many and varied. Thus it is time to move beyond the limits of well-rehearsed debates such as "large versus small scale" and "can organic agriculture feed the world?" to answer a much more discerning question: How can we work both together and within our different worlds to produce enough food to feed everyone in a way that is sustainable, equitable, and resilient?[1]

The State of Agriculture Today

Over the past few decades, the world's focus on increasing water abstraction for irrigation, supporting only a handful of high-yielding crop varieties, using petrochemical fertilizers and pesticides, and pursuing other technological "fixes" as a means to increase productivity has become a dominant way of thinking about agriculture—so much so that it is termed "conventional agriculture." While yield increases have undoubtedly occurred, this has also had a host of unintended environmental effects—degradation of land and water resources, biodiversity loss, pollution, and greenhouse gas emissions, to name a few—as well as socioeconomic effects—increased inequality, marginalization of the poor and women, and loss of community and household resilience to climate and economic shocks. And although the world grows enough food to feed the current population, poverty and hunger persist.[2]

At the beginning of 2011, one in seven individuals worldwide was chronically undernourished. Limited income and production opportunities for the poor and lack of effective social safety nets mean that around 925 million

---

Monique Mikhail is sustainable agriculture policy adviser at Oxfam.

people routinely do not have access to enough to eat. Many of these poor people are small-holder farmers or rural wage earners who have insufficient resources to meet their food needs. These rural poor are facing new drivers of hunger, including food price volatility and unpredictable weather caused by global climate change.[3]

In addition, the global food system is both contributing to degradation of the natural resource base and being squeezed by competing demands on it. Water is one resource feeling the squeeze. Agriculture both affects and is affected by water resources, accounting for 70 percent of global fresh water use. Pollution, caused by leaching of fertilizers and pesticides, degrades the quality of both surface and groundwater sources. Saltwater intrusion caused by overpumping of groundwater has irreversibly damaged some water resources. Increased irrigation with groundwater has caused water tables to drop, diminishing the ability of aquifers to hold water. Yields have declined on irrigated areas that suffer from waterlogging and salinization.[4]

Dominant agricultural practices have converted natural habitats and encouraged a shift toward monoculture production systems of a handful of export crops, which in turn has driven the loss of 75 percent of plant genetic resources over the past century. Only about 150 plant species are now cultivated commercially worldwide. And roughly 24 percent of the global vegetated land area has already been affected by human-induced soil degradation, particularly through erosion. Further, extreme weather events linked to global climate change—such as heat waves, droughts, and floods—have already begun to increase, with serious impacts on crop production, harvests, and food distribution and contributing in many local and national settings to food price spikes.[5]

Exacerbating all these problems is the capture of large land areas by companies, investors, and food-insecure governments. Governments and elites in developing countries are selling off large areas, much of it land that is already inhabited. Since the food price crisis of 2008, there has been a massive increase in these land deals: in just one year, the land investments in Africa equaled those of the previous 22 years.[6]

Despite these mounting problems with the food system and the disproportionate impact on poor small-scale producers, funding for developing-country agriculture has dropped significantly over the past few decades. Indeed, between 1983 and 2006 the global share of official development assistance for agriculture declined by 77 percent to only 3.7 percent, while support for agriculture in industrial countries climbed to more than $250 billion a year.[7]

Behind these trends are vested interests that have heavily influenced the balance of power, resulting in broken aid promises, blocked land reform, rigged trade rules, subsidies for wealthy farmers, and corporate power. For example, the Alliance for Abundant Food and Energy, founded by ADM, Monsanto, DuPont, John Deere, and the Renewable Fuels Association, was part of the biofuel lobby that influenced mandates for biofuel content in gasoline and diesel, subsidies, and tax breaks. This resulted in increased food price volatility.[8]

The Key Role of Small-scale Producers in a Sustainable Food System

What is needed is a new lens with which to see the food system, retaining the sustainable, equitable, and resilient elements and adjusting the remainder. Instead of focusing on techno-fixes such as agrochemical application, a fundamental shift toward an ecological approach to agricultural systems is required in science, technology, policies, institutions, capacity development, and investment. There is a huge potential for low-input, agroecological farming techniques to raise yields, improve soil fertility, conserve natural resources, and reduce dependence on expensive inputs. Several expert agencies and studies have reviewed the evidence base of the success of these approaches and are now advocating them. For example, studies of the System of Rice Intensification developed to help smallholders boost productivity and reduce reliance on inputs found average yield increases across eight countries of 47 percent and average reductions in water use of 40 percent.[9]

Changes are required across the whole food system, including critical shifts in large-scale production. But most of the world's poor depend on local markets for their food security. Therefore, small-scale food producers in developing countries are critical to achieving food security for the poor through sustainable, equitable, and resilient agricultural approaches. To achieve the level of change needed, increasing the quantity and quality of investment in small-scale food production is desperately needed: estimates show that an additional $50 billion per year of public investment is needed to eliminate hunger by 2025.[10]

Almost 2 billion people are fed by produce from the 500 million small farms in developing countries. Yet it is these very same small-scale producers who are the most food-insecure. In fact, about 80 percent of hungry people live in rural areas. The small-scale food producers there have considerable room for yield improvements that could increase the food security of their communities. Lower yields on small-scale farms in poor countries are largely due to a disparity of access to markets, land, finance, infrastructure, and technologies—not to inefficiency. Investing in approaches that address the inequity of access between large- and small-scale farmers will increase smallholder production, thereby raising their incomes and creating more inclusive agricultural growth.[11]

On top of yield improvements, supporting

An SRI (System of Rice Intensification) instructor in Cambodia offers advice to a farmer on how to pull rice seedlings without damaging their roots.

small-scale food producers can build sustainability and resilience to climate shocks. When smallholders are able to produce more food through techniques that are better for the environment, they are less vulnerable to future climate and economic shocks. In northeast Thailand, for example, jasmine rice farmers have been adapting to increased drought due to climate change by developing innovative ways to use water resources to improve their yields and help them in the future when drought strikes. Investment in helping these farmers to share their innovations has also improved the resilience of many of their neighbors.[12]

At the macroeconomic level, history shows that investing in agriculture can have a large impact on poverty reduction, not just because of the importance of agriculture for food security but also because developing countries depend heavily on the agricultural sector within their economies. Thus agriculture can provide the greatest "growth spark" for developing countries. In fact, growth that comes from the agricultural sector, particularly small-scale production, has twice the effect on the poorest as growth from other sectors. Evidence from the development paths of many of today's wealthy countries shows a variety of examples and models of investment in agriculture.[13]

The milk at this school in Sri Lanka is provided by the Oxfam supported dairy co-op located just around the corner.

The Importance of Addressing Gender Inequalities

Greater possibilities arise when applying a "gender lens" to investing in smallholder agriculture. Many small-scale food producers and agricultural laborers are women. In parts of Africa, women conduct 60 percent of the harvesting and marketing activities, 80 percent of storage and transport, 90 percent of hoeing and weeding, and 100 percent of processing of basic foods. However, unequal gender relations and persistent biases in beliefs, policies, and practices result in gross inequities. Women producers are systematically excluded from decisionmaking and often lack access to land, water resources, credit, information, and extension services. In fact, women receive just 7 percent of total aid to agriculture, forestry, and fishing. If women had the same level of access as men, their farm output would increase by 20–30 percent and global hunger would decline by 12–17 percent, according to the U.N. Food and Agriculture Organization.

Further, studies have shown that when women have control over income in a household, the money is more likely to go toward improving family food consumption, child nutrition, education, and overall well-being.[14]

In Sri Lanka, Oxfam worked with the government to develop dairy cooperatives with over 1,500 women producers collecting, processing, marketing, and distributing milk products. Milk production per cow quadrupled, significantly increasing women's incomes. Women have also improved their access to credit and have influenced the government to provide veterinary services as well as an insurance and pension scheme and to purchase their milk for local schools.[15]

Gender equity at the household and community level can also be improved through investment in women's agricultural livelihoods. Research in Mali, Tanzania, and Ethiopia has shown that women smallholders' engagement in collective action in different agricultural subsectors gives them access to inputs and markets and also helps overcome wider social barriers and improves social status, partly due to their increased contribution to household expenditures. Outcomes can include better positions in the household, increased decisionmaking power, and greater respect from their husbands for their opinions.[16]

The Need for Better Access

Most investment needs for smallholders boil down to increased access—access to natural resources, knowledge and information, financial services, credit, policymaking processes, and basic rural services. Foremost for smallholders is access to land and water resources for production, yet this is becoming increasingly difficult. In many cases, as noted earlier, developing-country governments are almost giving away land in response to increasing demand from industrial and rapidly growing economies. Much of this land already "belongs" to smallholders. Yet land tenure often involves discrepancies between legal and customary rights—and smallholders usually come up short.

Internationally applicable standards on good governance of land tenure and natural resource management are needed. These must include respect for and protection of existing land use rights and verification that local rights-holders have given their free, prior, and informed consent before land deals are endorsed. Governments should consider a moratorium on land rights transfers until these standards are in place and enforced. Investors must also be responsible and respect existing rights, avoiding transfer of rights (including those under customary tenure) from small-scale food producers. Alternatively, they can engage smallholders through fair contracts. Further, other actors in the value chain—such as financiers of agriculture ventures, traders, and processors should take responsibility for actions within their sectors.

Although land and water rights are often tied, access to water resources is essential in its own right. To increase production on existing farmland, two solutions are inevitable: irrigation or increasing soil moisture through soil and water conservation efforts. Pressures on land and water can be reduced through practices that boost yields, use soils and water more sensitively, and reduce their reliance on inputs—practices such as low- or zero-till agriculture, agroforestry, intercropping, and the use of organic manures. Low-cost, appropriate irrigation systems are also incredibly useful for enhancing productivity with small amounts of water and for producing crops in the off-season, when they fetch a higher price.

Community-based integrated water resource management (IWRM) can help communities keep water use within ecological lim-its. For example, Oxfam and Karkara, its local partner in Niger, have been working with rural communities since 2009 to undertake dry-season irrigation through community-based IWRM in an effort to safeguard food security and improve people's health. Community-led monitoring has provided communities with a long-term outlook and enabled key water user groups to make collective decisions about water usage and daily abstraction. Hydrological monitoring and training has also encouraged participation from local governments and regulating water authorities, providing a useful long-term external support platform for communities.[17]

While access to resources is critical, knowledge and information—about appropriate practices, weather, and pricing, for instance—can increase productivity and improve the sustainability of production. Small-scale food producers have invaluable indigenous knowledge and experience in managing climate variability and the conditions specific to their area. But they lack access to other forms of knowledge that could help them improve the productivity, sustainability, and resilience of their farms.

Agroecological practices are particularly knowledge-intensive, which begs the question: How is information delivered and to whom? Innovative efforts are being made to use information technologies—such as broadcasting crop prices over the radio and texting weather information to mobile phones—to fill some of the information transfer gaps. But this is not enough. Farmers learn best from hands-on interaction with other farmers and agricultural extension agents trained to share new techniques. Farmer-to-farmer training that Oxfam has undertaken in Honduras has increased the use of composting practices and the construction of living barriers alongside farming plots, which encourages farmers to stop burning stubble on their fields, builds up soil nutrients, and generates additional income.[18]

Unfortunately, along with the general decline in development support for agriculture over the past few decades, agricultural extension has been slashed. Yet many European countries, the United States, and Japan owe much of their agricultural productivity gains to periods of emphasis on solid extension services. For example, during its greatest period of growth, Japan had an extension worker for every village (about 100 households). Rebuilding these networks and developing their capacity to train smallholders on new agroecological techniques is vital. These modern extension services will also need to rebuild the link between research and extension that has diminished with declining funding.[19]

Global research and development (R&D) has been dominated by large companies that focus on technologies geared toward packages of products, such as Monsanto's Roundup herbicide and genetically modified Roundup Ready Soy, for large industrial farms instead of toward practices that are not easily sold but that can improve smallholder yields with less cost. R&D is also organized toward technology transfer, largely from research scientists of wealthy countries to farmers (mostly men) in developing countries. Although poverty alleviation is the primary goal, this approach can actually have negative impacts. Meanwhile, remarkable innovations made by smallholders in developing countries have been largely ignored by the broader development community. While participatory research has begun to grow, global R&D must be reformed to focus on technologies of practice, gender-based approaches, agroecological approaches, and a diversity of genetic resources, including important food staples that are not globally traded. Innovation and adaptation are iterative processes to be done in concert with smallholders and drawing on their social networks for the transfer of information.[20]

Markets are also difficult for smallholders, particularly women, to gain access to due to poor infrastructure and the private sector's reluctance to accommodate their different needs. Yet a number of existing approaches can assist smallholders with market access. At a minimum, farmers need price information, improved transportation infrastructure to allow them to reach markets, and storage infrastructure so they can store their crops and sell when prices are high. In the Amhara region of Ethiopia, a coalition of facilitating partners has developed the value chain for honey by providing producers (mostly women and landless people) with technology inputs and extension services, helping them to organize their production, and creating an enabling policy environment. Farmers who previously produced small quantities of low-quality honey have quadrupled their output and are now exporting certified organic honey to international markets.[21]

Organizing together can be critical for smallholder survival, particularly when farmers are interacting with competitive global market chains. Producer organizations can help smallholders in several ways:

- economies of scale to reduce transaction costs for buyers and make working with smallholders a more attractive proposition,
- greater bargaining power,
- better access to agricultural services, and
- a stronger political voice.

A producer organization in Mali, for instance, helped small-scale cotton producers overcome falling prices and increased privatization that had removed the state-owned cotton company and the inputs and training services it supplied. The producer organization replicated the lost state services for their members, helped them develop partnerships with lending institutions, and increased the participation of women in cotton co-ops.[22]

Unfortunately, producer organizations sometimes struggle to get governments to recognize them. For example, there is no law in Armenia to define and regulate a cooperative, making it impossible to form one. Strengthening and changing legislation to recognize producer organizations will give smallholders access to market functions and a political voice. In Indonesia, the local authorities in West Papua allocated resources to the vanilla sector after organized farmers convinced them of the potential for economic development and poverty reduction. Additionally, connecting producer organizations with the private sector has led to market access for smallholders in many of Oxfam's programs. In Sri Lanka, Oxfam's work with Plenty Foods to integrate 1,500 farmers into its supply chain has improved smallholders' access to land, credit, technical support, and markets, thereby increasing their incomes. In this win-win situation, the company registered a 30 percent growth over four years.[23]

Beyond access to markets, smallholders need access to financial services to manage risk and invest in their farms. Unlike large-scale producers, smallholders do not have the safety net necessary to take risks on new technologies or practices, as it can make the difference between feeding their families and going hungry. Better weather information and data, storage infrastructure, and access to insurance are all ways to help smallholders manage risks and invest in their farms. Financial tools like weather insurance can protect against revenue and yield losses. These types of insurance policies can also be packaged with credit and inputs like improved seeds or cash transfers for labor. In Tigray, Ethiopia, Oxfam piloted a Rural Resilience Initiative that had four main pieces: reduced risks from climate change through improved resource management, access to credit, savings to provide a buffer in difficult periods, and weather insurance offered through local companies to guarantee even the most marginalized farmers some income if bad weather ruined their harvests. The poorest farmers pay their premium through working on community projects like tree planting,

composting, and building irrigation systems. In the first three years of the program, the number of households with insurance climbed from 200 to 13,000.[24]

It is still generally extremely difficult for smallholders, particularly the most marginalized groups and women, to obtain financial services, even micro-credit. For example, women farmers receive just 10 percent of loans granted to smallholders and below 1 percent of total agricultural credit. To overcome these obstacles, Oxfam is supporting rotating savings and credit associations for women in West Africa, East Asia, and Central America that have shown great success in low-cost replicability, confidence building, and the creation of new opportunities.[25]

Moving Forward

Without government intervention to direct a more rapid transition, markets and the vested interests that govern them will not lead us toward a sustainable agriculture future. Clear global commitments and frameworks coupled with effective national and regional policies are necessary.

Yet there is no perfect blueprint for sustainable agriculture globally. Each agroecological zone and sociopolitical situation requires somewhat different policies in order to create an environment in which small-scale food producers can improve their livelihoods while maintaining environmental services. Appropriate solutions should be determined through livelihood, environmental, and political analyses that are context-specific. Yet there are some basic tenets and stakeholder roles that any agriculture and food security plan should contain in order to ensure the rights of small-scale producers are maintained and environmental sustainability is taken into account. Several practical approaches exist, although they are currently dispersed and operating on a relatively small scale. These approaches need to be scaled up through efforts at the local, national, and global levels.

Perhaps the largest investments in food production are made by food producers themselves. These should be supported and complemented by investments by national governments, international research institutions, the private sector, and donor governments to fill the current wide gaps.

National governments in developing countries must give priority to investments in key public goods such as capacity building, infrastructure, and research systems. Especially critical are sound investments in responsive and participatory extension services for small-scale food producers that improve the transfer of knowledge (including traditional knowledge), especially on natural resource management. In addition, improving food security will require public spending on basic education and social services designed to meet the needs of women. Special attention should be focused on service provision in marginal areas where small-scale food producers are most vulnerable to weather-related shocks and disasters and where they face land degradation and water scarcity.

International research institutions must incorporate small-scale food producer innovations as essential inputs to an iterative and inclusive research process. They should also recognize, support, and build on farmer networks to improve uptake of appropriate technologies and practices. R&D activities should emphasize developing technologies of practice over products, agroecological approaches that emphasize environmental sustainability, diversity of genetic resources, and approaches that are adapted to the specific needs and constraints of women food producers. R&D should also support small-scale food producers in their adaptation to climate change.

Private-sector actors must develop and adhere to equitable principles of partnership and engagement that integrate small-scale food producers into value chains under fair terms, that share and manage risks associated with agriculture, that implement inclusive practices for access to credit and technical support that addresses the key constraints of small-scale food producers, and that ensure that investments contribute to and do not harm the food security of families, communities, and countries. Private-sector actors also need to ensure that their actions and investments protect and restore natural resources.

Donors should reaffirm a strong commitment to meet their pledge in 2009 to invest $20 billion over three years to tackle food insecurity in developing countries. These commitments should be measured and reported against poverty, food security, and nutrition outcomes. In addition, donor coordination, alignment, and support for country-led plans are needed. Investments must be predictable, transparent, untied, and channeled through budget support where appropriate. Beyond the three-year commitment term, a new multilateral food security framework is needed, particularly focusing on country and regional plans that are developed in a transparent and inclusive manner and that support the specific actions just described.[26]

With this focus on increasing the quantity and quality of investment in small-scale food production, applying a gender lens to investments, and addressing issues of access alongside intentional cultivation of a more agroecological approach to farming, we could indeed grow a food system that achieves food security while sustaining Earth's systems and maintaining ecosystem diversity.

CHAPTER 14

Food Security and Equity in a Climate-Constrained World

Mia MacDonald

In August 2011 the first outpost of the American fast-food chain KFC, famous for its fried chicken, opened in Nairobi, and hundreds of Kenyans waited in line for more than an hour to get in. KFC could only enter the market, its executives say, after resolving supply chain issues. KFC's chicken will come from Kenchic, a large domestic producer with industrial-style poultry sheds in Mlolongo, not far from Nairobi's city center.[1]

In Shanghai, during a discussion of livestock and climate change, a professor noted that "it's a global, capitalist economic system. We need to [adopt the intensive model of meat production]. If we don't, we'll lose out."[2]

And in India, still home to the world's largest population of vegetarians (although omnivores now form a majority), the recently established National Meat and Poultry Processing Board declares on its website: "As the country's livestock industry is changing, India attempts to become a key player in the global meat market."[3]

As these few examples illustrate, in recent decades a "livestock revolution" has spread to Asia, Latin America, and, to a lesser extent, Africa. Until relatively recently in human history, regular consumption of meat was limited to the wealthy elite. Most people ate meat only on special occasions, since their animals—cows, sheep, goats, pigs, chickens, and others—had more value alive than dead. But over the past 60 years or so, vast changes in agricultural production in industrial countries, including the use of large, confined, factory-like facilities that house thousands of animals, have made meat, dairy products, and eggs much more widely available and affordable.

Today in developing countries, and in their cities in particular, animal products have become part and parcel of a growing number of people's daily meals. Eating meat now often represents prosperity, independence, or modernity in a world where western-style consumption has set an international standard. People in industrial regions still eat much more meat than people in developing countries do: on average, 80 kilograms a year compared with 32 kilograms. But the gap is narrowing, and more than half the world's meat is now produced and consumed in developing regions.[4]

Since the 1970s, global meat production has

Mia MacDonald is executive director of Brighter Green, a nonprofit public policy action tank based in New York City that focuses on the environment, animals, and sustainability, and a Senior Fellow at Worldwatch Institute.

grown nearly threefold—and it has increased 20 percent just since 2000. Each year, more than 60 billion land animals are used in meat, egg, and dairy production around the world. If current trends continue, by 2050 the global livestock population could exceed 100 billion—more than 10 times the expected human population at that point.[5]

Industrial-scale livestock production has allowed farmers to raise enormous numbers of animals. Yet these facilities are far more factory than farm. Animals—hundreds, thousands, or even tens of thousands in the case of chickens—are confined in small pens, cages, sheds, or stalls in indoor sheds. They have been almost wholly removed from the land: they lack access to pasture, the remains of the harvest, fresh air, and sunlight. Cows raised for beef may graze for some months, but they spend much of their lives in dirt "finishing" feedlots.

An intensive "broiler" chicken facility near New Delhi, India

Domestic animals in industrial systems have been bred intensively to ensure large quantities of meat, milk, or eggs and to grow quickly on feed in which corn and soy feature prominently (antibiotics and growth-promoting hormones may also be present). An industrial pig is ready for slaughter at six months; a standard meat chicken in six weeks. Intensive facilities for pigs and poultry are those that are expanding most rapidly around the world.

Into the Dark

Chicken production in India once relied on small, backyard flocks raised by individual farmers, many of them women. Today 90 percent of the more than 2 billion chickens in India that come to market each year arrive from industrial facilities. In fact, India—where ethical vegetarianism has a long history—is now the world's fifth largest producer of poultry meat. In 2010, it was the world's fastest-growing poultry market, outpacing Brazil, China, the United States, the European Union, and Thailand.[6]

Advocates of factory farming argue that this form of agriculture is necessary if the world's population is to be fed. But a growing number of critics—within civil society and policymaking bodies in Europe, the United States, and the United Nations, as well as in countries going through the livestock revolution now—argue the opposite: that this model risks creating worse ecological and climatic conditions, greater food insecurity, and poorer public health than it can remedy.

A multiyear study by the Pew Commission on Industrial Farm Animal Production concluded that factory-farm facilities in the United States "have produced an expanding array of deleterious environmental effects on local and regional water, air, and soil resources. Those effects impose costs on the society at large that are not 'internalized' in the price paid at the retail counter for meat, poultry, dairy, or egg products….The volumes of manure are so large that traditional land disposal methods can be impractical and environmentally threatening. Excess nutrients in manure contaminate surface and groundwater resources."[7]

Similar realities are being seen in fast-emerging economies, too. In China, for instance, "domestic animal and poultry waste has become a major source of environmental pollution," according to Wu Weixiang, associate professor at Zhejiang University's College of Agriculture. Studies by Xu Cheng, a professor at China Agricultural University, concluded that China's livestock produce 2.7 billion tons of manure a year, nearly three and a half times the amount of industrial solid wastes. Of the 20,000 large- and medium-sized livestock operations that Xu estimated existed in China in 2007, a mere 3 percent had facilities to treat animal wastes. Rapid expansion of fish farming is also creating environmental problems. (See Box 14–1.)[8]

Feeding People or Feeding Animals

"I eat sausage in the morning, a meat dish and a vegetable dish for lunch, and the same for dinner. If there's no meat, I won't feel full, but if there's no vegetable, no problem," says Beijing student Guo Meng. Meng is not alone. Many in China's middle class now eat meat every day, sometimes at every meal. The percentage of energy from fat (from both animal and vegetable sources, including oils) in the average Chinese diet increased by 10 percent between 1996 and 2006. By 2006, about 60 million Chinese were obese, according to the State Food and Nutrition Consultant Committee. Diet-related chronic diseases now kill more Chinese than anything else.[9]

Eating more animal products, along with more sugar, salt, and processed and fried foods, is just one consequence of the rapid globalization of the western diet. Such shifts, seen most markedly among urban populations in developing countries, also mean that people eat fewer vegetables, fruits, and whole grains. These dietary changes—described by Dr. Frank Hu of the Harvard School of Public Health as "the most rapid and dramatic in the course of human history"—along with increasingly sedentary lifestyles, have led to growing concern about the rising global incidence of non-communicable disease.[10]

In India, rates of obesity, diabetes, and heart disease are increasing rapidly, potentially worsening social inequalities as the health system struggles to treat conditions like malnutrition, infant and maternal mortality, tuberculosis, and HIV/AIDS. Some 50 million Indians have diabetes, according to the International Diabetes Foundation, a number expected to reach 87 million by 2050. The Economist Intelligence Unit calculated that India paid the highest price of any country for diabetes: 2.1 percent of gross domestic product a year, when formal medical care and lost productivity are combined. But even as more Indians eat higher up the food chain, with sometimes undesirable health outcomes, undernutrition remains stubborn and persistent. Forty-four percent of children in India under the age of five are malnourished.[11]

Equity in terms of the use of natural resources is also put in the balance. Although factory farms and feedlots appear to use less land, since the animals are kept in confined areas (thereby avoiding obvious deforestation or soil erosion), their enormous feed requirements must be met by using other land. Through the current food system, "you could even feed 8 billion [people], maybe you could feed 9 billion," U.N. Population Fund advisor Michael Herrmann says, but he adds that "a large share of the food we produce does not actually end up as food on our plates." Corn and soybeans are fed to animals; they, in turn, are fed to us.[12]

An inherent and troubling inefficiency exists. Between two and five times as much grain is required to realize the same number of calories through livestock as through grain eaten by people directly (and up to 10 times as much for industrially produced beef), according to research by Rosamond Naylor of Stanford University.[13]

China apportions more than half of its corn

Box 14–1. Aquaculture's Costs and Benefits

During the 1970s total world aquaculture production amounted to less than 5 million tons—just 4 percent of total fish harvest. But by 2009 it reached 55.1 million tons—38 percent of the total harvest. The rapid development of aquaculture has filled the shortfall in the global wild catch market, which peaked at about 90 million tons in the mid-1990s. While bringing some benefits, especially in developing countries—by offering year-round employment, providing valuable sources of protein, and contributing to the national economy—aquaculture also has serious ecological side effects.

According to the Food and Agriculture Organization (FAO), the world will need to produce an additional 40–60 million tons of fish per year by 2020 just to maintain fish consumption per person at current levels (17 kilograms a year). This is about 19 percent more than 2009 production—and much of this increase (85 percent) is projected to come from aquaculture.

Nearly 60 percent of the world's fishers are small-scale commercial or subsistence individuals, mainly in developing countries, which can contribute significantly to food security and poverty reduction in these nations. But with shifting trends, employment in wild catch fisheries is stagnating or declining worldwide, while jobs in aquaculture—at almost 11 million people in 2008 —are growing substantially. Fish farmers in China have experienced the greatest increase: 189 percent between 1990 and 2008.

Pressure to be more competitive and respond to market forces has increased the intensification of aquaculture production, with major environmental impacts: physical degradation of freshwater and coastal habitats through soil impoverishment, conversion of mangrove forests and the destruction of wetlands, salinization of agricultural and drinking water supplies, and land subsidence due to groundwater abstraction. Intensification contributes to eutrophication through the discharge of unused animal food and the excreta of fish and chemicals. For every ton of fish, aquaculture operations produce 42–66 kilograms of nitrogen waste and 7.2–10.5 kilograms of phosphorus waste. Lake Taihu—the third largest freshwater lake in China, with a total water area of 2,338 square kilometers—is used for aquaculture farming. After only a few years, the phytoplankton and nutrients have increased by several factors, and eutrophication has led to a measured deficit in oxygen in surface sediments.

FAO has identified the ecosystem approach as a fundamental step toward effective environmental management and sustainable resource use in fish farming. In Thailand, the top exporter of fisheries products since 1993, regulators are paying more attention to sustainable management of fisheries. The use of mangrove areas for shrimp farming is currently restricted to designated areas by permission of the Department of Forestry, which controls the mangrove forest. Regulations have been set on farm size and wastewater treatment.

—*Trine S. Jensen and Eirini Glyki*
Worldwatch Institute Europe
Source: See endnote 8.

supply (domestically produced and imported) to livestock feed, up from about 25 percent in 1980. In order to ensure food security, China is buying more commodities on global markets—principally soy, but also corn. And it is leasing land in other countries to grow food for its people as well as its farm animals.[14]

Globally, 85 percent of soybeans are processed into meal and oil, and 90 percent of the meal is used to manufacture animal feed.

China now purchases more than half of the soybeans traded on world markets. Its growing demand has been met in large measure by the expansion of soy acreage in Brazil. (See Box 14–2.) China purchases more than 40 percent of Brazil's soy.[15]

A lack of land for agricultural expansion also has led Indian agricultural producers to secure long-term leases or to purchase land outright outside the country's borders. Indian agribusinesses have formalized agreements in Kenya, Madagascar, Mozambique, Senegal, and Ethiopia to grow and export back to India rice, sugarcane, palm oil, lentils, vegetables, and corn (the last for feed).[16]

In recent years, the ethics of using corn, palm oil, and sugarcane to produce biofuels has come under more scrutiny, justifiably, for its potentially harmful impacts on global food prices, world hunger, and the environment. Yet in the year 2007/08, just 4 percent of global production (100 million tons) of cereals was used for biofuels, of which 95 million tons was corn. Compare that to the 35 percent (756 million tons) of cereals that was used for animal feed. In 2007, just 12 percent of the world's corn was used for ethanol, while 60 percent was used to produce animal feed. Between 2010/11 and 2011/12, the amount of world cereal production for use in animal feed rose by 1.5 percent, a higher percentage than for other uses (such as for food for humans or fuel).[17]

Water Pressure

Ethiopia, home to Africa's largest livestock population and with a recent history of persistent food insecurity, may face a stark choice in coming decades: use available water and land resources—already under considerable pressure from the effects of soil erosion, recurring drought, and deforestation—to grow enough food for a fast-rising human population or grain for cattle raised in feedlots and chickens in broiler and layer sheds.[18]

Intensification of animal agriculture means that "the livestock sector enters into more and direct competition for scarce land, water, and other natural resources," according to FAO. This has enormous consequences for equity and sustainability—and for achieving broad-based prosperity for the world's people.[19]

China and India, with less than 10 percent of the world's water between them, are home to one third of the world's people. According to a 2009 U.N. report on water and development, shortfalls due to climate change, urbanization, population growth, and the needs of agriculture and food production pose significant challenges to continued rapid economic growth in coming decades across Asia.[20]

The U.N. report noted concerns about rising consumption in emerging economies of meat, eggs, and dairy products, which are far more water-intensive than "the simpler diets they are replacing." Each ton of beef requires 16,000 cubic meters (4.2 million gallons) of water.[21]

A UNESCO study calculated that 29 percent of the "water footprint" of the world's agricultural sector is due to production of animal products. In importing Brazilian soybeans, it is worth noting, China is also importing "virtual water"—up to 14 percent of what it needs, according to one analysis.[22]

Vegetarian diets, like those still popular among millions of Indians, require an average of 2.6 cubic meters of water for each person each day, according to Shama Perveen at the Indian Institute of Management in Kolkata. The diet of the average person in the United States, in contrast, which contains much higher quantities of poultry, beef, and dairy, needs more than twice as much daily water: 5.4 cubic meters.[23]

Climate Change Impacts

Expanding global meat, egg, and dairy production has a direct relationship to global climate change. Approximately 18 percent of

Box 14–2. The Changing Nature of Agriculture in Brazil

Brazil, the world's most biologically diverse nation, leads the world in exports of beef and veal. It is also the top global exporter of poultry meat, the fourth largest of pork, and the second biggest of soy.

To keep pace with international demand and rising domestic consumption, Brazil's livestock producers have added animals, facilities, and processing and transport capacity. Many large-scale operations are located in the south, near ample supplies of feed ingredients (soybeans and corn).

While most of Brazil's nearly 200 million cattle are still free-ranging, much of the pasture they graze on has been carved from areas of extreme biological diversity, specifically in the Amazon rainforest and the cerrado—the Brazilian savannah. Both regions are also demarcated by a patchwork of large, straight-edged fields, akin to those in the U.S. farm belt, planted with row after row of soybeans.

Genetically modified (GM) soy has a growing foothold. Two thirds of Brazil's 2009–10 soy crop was Roundup Ready, a breed genetically engineered to be resistant to the herbicide Roundup and sold by U.S.-based agroscience corporation Monsanto. Cultivation of GM corn

in Brazil is also rising and accounts for about 40 percent of all corn acreage in the country.

Brazilian civil society groups have expressed concern about the environmental, public health, and economic impacts of GM soy. The small farmer is "forced to buy seeds and the package [of] inputs that increase the costs of production.... His profit is going to the corporations," says Tatiana de Carvalho, an agronomist and consultant for Greenpeace in Brazil.

In industrial animal agriculture, control over production is by and large concentrated among a small number of powerful agribusinesses or large landowners. In Brazil, smaller-scale, independent farmers have been pushed to the margins of the agricultural economy. Some have become *integrados*, or contract farmers, for large conglomerates. Others, lacking the capital to become contractors, unwilling to give up their autonomy, or facing harassment from large landowners, have moved to a city. This may end in unemployment or part-time work and potentially hunger too, since they no longer have land on which to grow food.

Source: See endnote 15.

global greenhouse gas (GHG) emissions can be traced to the livestock sector, according to FAO (9 percent of global carbon dioxide emissions, 37 percent of global methane emissions, and 65 percent of global nitrous oxide emissions). An analysis by current and former World Bank environmental specialists put the figure at 51 percent of total global GHGs.[24]

In Brazil, GHG emissions from agriculture increased 41 percent between 1990 and 2005. It is estimated that 75 percent of Brazil's GHGs are due to deforestation and changes in how land is used, which clear the way for livestock and crops. A 2009 estimate (deemed conservative) concluded that fully half of

Brazil's GHG emissions between 2003 and 2008 arose from the cattle sector alone.[25]

In 2009, scientists at India's Space Applications Centre conducted the first national study of emissions of methane from India's nearly half-billion cows, buffalo, sheep, and goats. They found that the emissions had risen almost 20 percent between 1994 and 2003, to 11.75 million tons annually. The figure is undoubtedly higher now: between 2003 and 2007 India's population of cows and buffalo increased by 21 million.[26]

The amount of methane released is directly related to how and how much ruminants are fed. Indigenous cattle emit less methane than so-called high-yielding breeds of imported cattle, but the latter are gaining in popularity among large-scale Indian dairy operators. One "mega dairy" project on the drawing board in India would eventually contain up to 40,000 cows producing milk, many slated to be imported, high-yielding breeds, all eating exclusively grain-based feed.[27]

Plant-Based Foods and a Well-Fed World

To feed all the world's people by 2050, given rising populations and incomes, food production must increase by 70 percent, according to a recent FAO report. But at the same time, the report noted, 25 percent of the world's land is degraded, and water is becoming increasingly scarce and polluted, both above and below ground. Moreover, FAO warns, as global warming shifts climate patterns, competition will become "pervasive" for water and land, including within the agriculture sector between livestock, staple crops, non-food crops, and biofuels. In this scenario, large-scale livestock production and ever-rising global consumption of animal products represent an enormous challenge rather than a solution.[28]

The U.N. Environment Programme (UNEP), in assessing the impacts of how materials are produced and used, concludes that "more than fossil fuels, agricultural activities directly influence ecosystems by occupying large land areas and using huge quantities of water." Unlike shifting away from using fossil fuels, however, there is no alternative to eating. So, UNEP adds, "a substantial reduction of impacts would only be possible with a substantial worldwide diet change, away from animal products." Such a change would also require a serious rethinking of factory farming and the role of meat, dairy, and eggs in achieving global food security, not just a tinkering with the parts.[29]

This rethinking will require concerted,

creative actions by governments as well as civil society.

Governments, for example, should ensure that water pollution and land degradation, deforestation, harm to or destruction of ecosystems and biodiversity, and greenhouse gas emissions, are no longer "external" to the livestock industry's balance sheet, but rather priced fairly and fully paid for. This would entail setting prices for ecological services and GHGs at market rates or mandating mitigation technologies for facilities already operating or on the drawing board.

Healthy spinach in a school garden, Zimbabwe

Governments, in collaboration with civil society, ought to lay out alternatives to the industrial agricultural system that would be better for the climate, the environment, family farmers, and food and income equality. Shifting the locus of investment and policies away from monocultures of livestock and crops and toward non-industrial-scale farmers and an array of produce grown with more-sustainable agricultural practices like agroecology would be essential. Systems of land tenure ought to promote protection of forests, grasslands, and other ecosystems—in other words, conservation for carbon sequestration.

In addition, governments ought to provide incentives to promote cultivation of foods that provide key nutrients, like leafy greens and pulses, which require less water than soybeans or feed grains and may be more resilient to the anticipated effects of climate change. They also need to ensure equitable access to such foods. These measures do not have to consign farmers to operating at subsistence levels on small plots of land; they can include medium-sized enterprises and mechanisms for sharing agricultural risks and returns, such as cooperatives.

Donor agencies, governments, and civil society should collaborate on large-scale ecosystem restoration projects that would revitalize overgrazed and overharvested lands and create new opportunities for increased production of nutritious food (as well as the regrowth of forests and vegetation that help ensure stable rainfall patterns). These restoration efforts could also create new jobs or livelihoods.

Governments should pass legislation on animal welfare that would end the abuses and cruelty inherent in factory-style production facilities. In many societies, such laws would reflect national cultural heritage and values that protected animals and habitats for generations, as well as the country's own constitution. India's constitution, for example, includes the duty of every citizen to have compassion for living creatures. Kenya's 2010 constitution has a provision on animal welfare, and Ecuador's 2008 constitution enshrines the rights of nature.[30]

The government and civil society—groups on the environment, food security, antipoverty, small farmers, women, and animal welfare—should participate in serious, wideranging national and state-level dialogues on food production and food access, livestock, sustainability, and equity that would help put national and regional government policies in place. A forum like this could also help increase the public's awareness of this complex and critical set of issues.

With the participation of a range of civil society groups, governments ought to launch national public education efforts to encourage adults and children to eat more healthily. The starting point should be traditional, largely plant-based regional and national cuisines. The objective would be twofold: to advance food security and sustainability and to reverse upward trends in rates of chronic diet-related conditions like obesity, diabetes, hypertension, and some cancers. Governments could also support campaigns like Meatless (or Meat Free) Monday, which ask individuals and institutions to not eat meat one day a week for their health and the environment. Meatless Monday efforts have taken root in a growing number of U.S. and European cities, as well as in Cape Town, South Africa, and several Brazilian municipalities. Industrial nations, where factory farming and high animal-products consumption are entrenched, could take the lead here—providing an example for the rest of the world.[31]

Finally, nongovernmental and community-based organizations in all countries that work on issues of climate change, food security or sovereignty, food safety, resource use, rural livelihoods, and animal welfare should exchange experiences, insights, and information with their counterparts elsewhere.

It is increasingly clear that food and agricultural production need to be central to sustainable development and climate policy, including any agreements on stemming deforestation, transferring green technologies, strengthening green economies, and funding poorer countries' adaptation to global warming. Continuing to overlook animal agriculture in these arenas means forfeiting a crucial opportunity to create a sustainable, equitable, efficient, humane, and climate-compatible food system.

CHAPTER 15

Biodiversity: Combating the Sixth Mass Extinction

Bo Normander

At the 1992 Earth Summit in Rio, world leaders made a collective commitment to preserve Earth's biological resources by agreeing to the Convention on Biological Diversity (CBD). Since then, however, most politicians have failed to protect nature and the world has witnessed—with but a few positive examples—a dramatic and continual loss of biodiversity. Not only have exceptional mammals such as the Western Black Rhino, the Caspian Tiger, and the Pyrenean Ibex Goat gone extinct, but an alarming number of animals, insects, and plants are now on the edge of extinction. It may not be long before the classic "poster species" such as the panda bear, the tiger, and the Baiji river dolphin disappear in the wild—kept alive only in public zoos by expensive breeding programs.[1]

The *Red List of Threatened Species* prepared by the International Union for Conservation of Nature clearly shows the alarming trends for biodiversity, measuring seven categories of extinction risk. Nearly one fifth of the almost 35,000 vertebrate species evaluated so far are classified as "threatened," ranging from 13 percent of birds to 41 percent of amphibians. (See Figure 15–1.) From 1980 to 2008, on average 52 species a year moved one category closer to extinction. Of all groups evaluated, cycads and sturgeon have the highest proportion of threatened species, at 64 percent and 85 percent respectively. Cycads (palm-like plants) are found in many tropical and subtropical areas and are the oldest seed plants in the world. The main threats they face are habitat disturbance, loss of habitat due to urbanization, and illegal removal by collectors. Sturgeon are also ancient species, among the oldest families of fish in the world. The Beluga sturgeon of the Caspian Sea produce roe that can be worth up to $10,000 per kilogram for their use as black caviar. Caviar demand has caused severe overexploitation of sturgeon populations throughout Europe and Asia.[2]

A second revealing measure of biodiversity loss is the Living Planet Index, which is based on monitoring populations of over 2,500 vertebrate species. It reflects a similar negative trend, as biodiversity declined by 12 percent at the global scale and by 30 percent in the tropical regions since 1992. (See Figure 15–2.) In other words, the rate at which species are becoming extinct is estimated to be up to 1,000 times higher today than in pre-industrial times. Scientists have called this the sixth mass extinction in Earth's history—and the only

---

Bo Normander is the director of Worldwatch Institute Europe.

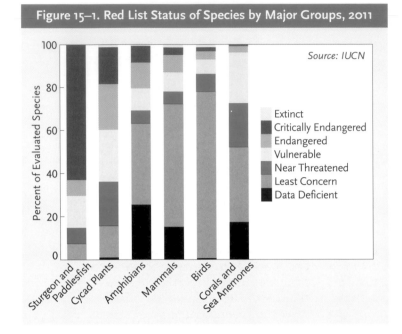

Figure 15–1. Red List Status of Species by Major Groups, 2011

Source: IUCN

Legend:
- Extinct
- Critically Endangered
- Endangered
- Vulnerable
- Near Threatened
- Least Concern
- Data Deficient

y-axis: Percent of Evaluated Species

x-axis categories: Sturgeon and Paddlefish, Cycad Plants, Amphibians, Mammals, Birds, Corals and Sea Anemones

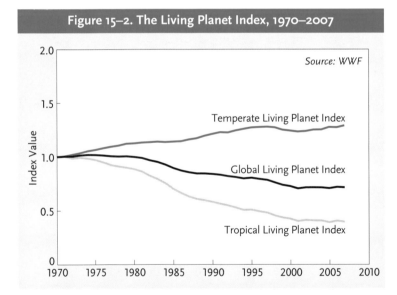

Figure 15–2. The Living Planet Index, 1970–2007

Source: WWF

y-axis: Index Value

Temperate Living Planet Index
Global Living Planet Index
Tropical Living Planet Index

x-axis: 1970, 1975, 1980, 1985, 1990, 1995, 2000, 2005, 2010

What is the cause of this biological tragedy? The answer is simply human intervention. The CBD Secretariat points to five principal pressures that are driving biodiversity loss: habitat change, overexploitation, pollution, invasive alien species, and climate change. Over the last few decades humans have changed ecosystems to a degree that has not previously been seen. To sustain economic growth and the increasing demand for food, resources, and space, large parts of the planet's natural areas have been transformed into cultivated systems such as agriculture and plantations and into built environment. In 2005 the Millennium Ecosystem Assessment assessed that 15 out of 24 "ecosystem services" are in decline, including freshwater resources, marine fish populations, and access to clean air and clean water. (See Chapter 16.)[4]

Why Biodiversity Matters

So evidence is piling up and the message is clear: biodiversity is being lost at all scales. But why should anyone be concerned about biodiversity? As long as the world can produce enough food and obtain enough wood, fuel, and other

one caused by a living creature: humans. The other five mass extinctions happened long ago, with the end of the Cretaceous period 65 million years ago, which killed the dinosaurs, as the latest and most famous.[3]

resources from forests, farmland, and the oceans, why worry about a few thousand rare species that no one has ever heard about? For many people this is a relevant question. They do not fully understand or appreciate the importance of biodiversity—or even the meaning of the word. In a European survey from 2010, two thirds of those interviewed said they had heard of biodiversity but only 38 percent could explain what it meant. Nonetheless, when informed about the meaning, 85 percent replied that biodiversity loss is a very or fairly serious problem.[5]

A straightforward schoolbook definition of biodiversity (or biological diversity) might be the variation of life at all levels of biological organization. The most widely accepted definition is probably the one adopted by the CBD in 1992: biodiversity is "the variability among living organisms from all sources including, inter alia, terrestrial, marine and other aquatic ecosystems and the ecological complexes of which they are part; this includes diversity within species, between species and of ecosystems."[6]

The CBD definition is broad in the sense that it addresses not only the diversity of all living organisms but also the diversity of ecological complexes of which these are part. So biodiversity is not just about fighting whale hunting, as portrayed in the famous *Free Willy* films, or saving the panda, as symbolized in WWF campaigns. It is about preserving all life in all its forms.

To understand the importance of biodiversity in a given habitat or ecosystem, think of biodiversity as a gigantic house of cards, with each card representing a single species or ecosystem function. A few cards may be removed without much happening to the house. But if the wrong card is pulled out the whole house will collapse. In the same way, biodiversity is a complex system of literally millions of different species—from tiny microorganisms up through the hierarchy to the top predators—interlinked through food webs, pollination, predation, symbioses, antibioses, and many other chemical and biological interactions, many of which are not even known. Damaging part of the system—wiping out a few key species, for instance—

may lead to the collapse of the whole system.

Extensive deforestation on Easter Island dating back to the fifteenth and sixteenth centuries, for example, has caused the extinction of native trees, plants, insects, and all native bird species, as well as irreversible devastation of the whole ecosystem, leading to today's problems of heavy soil erosion and drought. Similarly, the introduction of non-native species can be fatal to ecosystems, as exemplified by the famous case of rabbits in Australia. Since their introduction by European settlers in 1859, the effect of rabbits on the ecology of Australia has been devastating. They are responsible for the major decline and extinction of many native mammals and plants. They are also responsible for serious erosion problems, as they eat native plants, leaving the topsoil exposed and vulnerable.[7]

Beyond the serious negative consequences of biodiversity loss, from an ethical standpoint human beings do not possess the right to decide or judge which species will survive and which will not. All species are equal, and humans have no right to eliminate living organisms by the thousands. Preserving biodiversity is also vital from a more anthropocentric point of view: it is not just about the human desire to be in and enjoy exciting and diverse nature, it is about the intact ecosystems that meet people's basic needs for food, clean water, medicine, fuel, biological materials, and so on.

Biodiversity is invaluable and cannot be truly measured in monetary terms. Even so, a recent U.N. Environment Programme (UNEP) study attempted an economic evaluation of a predefined characteristic of biodiversity. It suggested that if just 0.5 percent of the gross world product were invested in greening natural capital sectors (forestry, agriculture, fresh water, and fisheries), it would help create new jobs and economic wealth while at the same time minimizing the risks of climate change, greater water scarcity, and the loss of ecosystem services. In other words, preserving Earth's biological diversity is a fundamental step on the path toward achieving economic prosperity.[8]

Unmet Policy Targets

Despite the argument that preserving biodiversity is essential to human wealth, real political efforts to do so are yet to be seen. In 2002, parties to the CBD committed themselves "to achieve by 2010 a significant reduction of the current rate of biodiversity loss." Eight years later the same parties met in Nagoya, Japan, and concluded that the target had not been met—not globally, not regionally, and not nationally. Consequently, the target was renewed with the adoption of the Strategic Plan for Biodiversity 2011–2020 with 20 new headline targets—called the Aichi Biodiversity Targets—to "take effective and urgent action to halt the loss of biodiversity in order to ensure that by 2020 ecosystems are resilient and continue to provide essential services."[9]

Stig Nygaard

Wildlife in a park on the outskirts of Copenhagen

Some of the Aichi targets are ambitious, others are less stringent, and some issues have been left unaddressed. For example, urban biodiversity has been entirely overlooked. (See Box 15–1.) But overall the Strategic Plan reflects a growing international acceptance of the importance of biodiversity. Yet this global agreement needs to be followed up by concrete and ambitious national action plans as well as by real integration of biodiversity values in all relevant policy areas, societal sectors, and national accounts. This will be the main challenge for governments in the years ahead. Unfortunately, the past has witnessed a great deal of failure in this field. The fact that almost all countries left the 2010 targets unmet in silence and without any consequences whatsoever reflects the fundamental lack of political will to act on the urgent need to save biodiversity.[10]

Biodiversity Loss versus Climate Change?

Widespread public awareness of the dangers of climate change emerged less than a decade ago, culminating in 2007 when the Nobel Peace Prize was awarded jointly to the Intergovernmental Panel on Climate Change (IPCC) and former U.S. vice president Al Gore for their efforts on this issue. Biodiversity loss has not yet achieved the same amount of attention as climate change, despite similarly dire consequences. In a 2009 study in *Nature*, scientists named biodiversity as the "planetary boundary" that humans have surpassed more than any other, underlining the urgency of combating its loss. Yet this issue does not have nearly the amount of scientific knowledge and consensus as the field of energy sources and climate change.[11]

In early 2011 governments decided to create the Intergovernmental Platform on Biodiversity and Ecosystem Services (IPBES). Like the IPCC, which was created in 1988, the IPBES should be an interface

Box 15–1. Urban Farmers Can Reduce Biodiversity Loss

Protecting biodiversity in urban areas has been recognized as an increasingly important issue. In part, this is a consequence of rapid urbanization. In 2009, for the first time in history, more than half the world's population was living in urban areas. Urban growth is projected to continue in the coming decades, although at a decreasing rate, meriting special attention in order to make life in cities more environmentally sustainable.

Urbanization has a negative overall impact on biodiversity and especially on native flora and fauna in areas under urban sprawl. But all native species do not suffer because of urbanization, and the species abundance and diversity in some areas and especially on the fringes of cities can be much higher than—although often very different from—the diversity of surrounding rural areas. In Denmark, a study found that the urban area of Copenhagen, with its parks, forests, lakes, beaches, wildlife refuges, and other green areas, hosts a wide variety of species and in fact is one of the richest localities of biodiversity in the country. While more than 60 percent of Denmark's land area is intensively farmed, leaving little room for biodiversity, pockets of rich nature remain in a number of semi-urban areas.

For decades nature has been forced out of cities. Even the 2020 Strategic Plan for Biodiversity does not address aspects of urban biodiversity. To counter this oversight, a number

of grassroots and municipal initiatives have emerged, such as urban beekeeper associations, rooftop farmers, and vertical garden projects in Amsterdam, Singapore, New York City, and a growing number of other cities. These initiatives can reverse biodiversity loss and encourage urban greening and agriculture; they can also provide a way to improve quality of life, nourishment, and the integration of nature into city life. Another example is window farming. In one case, over a year more than 13,000 people worldwide downloaded instructions for building windowfarms, growing their own fruits and vegetables, such as strawberries, tomatoes, and peppers, in window openings.

Urban farming and gardening is a way to help stop environmental destruction and the loss of biodiversity. As Jac Smit, founder and past president of The Urban Agriculture Network, pointed out, urban farming "creates green spaces, recycles waste, cuts down on traffic, provides employment, substitutes for imported high value goods, prevents erosion and is good for the microclimate." There are many urban land areas not in use today that can be turned into green spaces. To get started, local authorities should be required to provide information on land use in urban areas and to adopt favorable city planning so that people can create new green and diverse spaces.

Source: See endnote 10.

between the scientific community and policy-makers. But much greater resources will need to be invested in IPBES—an obvious task for the participants at the Rio+20 Conference in June 2012—if this body is to be as significant as the IPCC. The IPBES should gather all leading experts and scientists to address the most recent scientific, technical, and socio-economic information to help make the 2020

biodiversity targets workable and achievable, while also raising the world's attention to the issue of biodiversity change.[12]

A significant drawback to the Aichi biodiversity targets is that they carry relatively weak legal obligations, if any. In contrast, the Kyoto Protocol on climate change implies legal obligations and effectively becomes a contractual arrangement for the ratifying country. In addition, the Kyoto Protocol outlines concrete and measurable national targets, such as reduction goals for emissions of greenhouse gases, whereas the biodiversity targets are more imprecise, vague, and difficult to monitor.

Yet the Kyoto Protocol, while well intentioned, seems to be doomed to fail its reduction targets anyway. The IPBES should work fast to adopt a simple and accessible approach to aggregate data in order to set up national targets for protecting biodiversity. There is, of course, no conceivable indicator that can accurately reflect changes in biodiversity in different ecosystems at different spatial and temporal scales because of the inherent complexity of habitats within ecosystems. But the IPBES could define a subset of indicators that can reflect balanced national assessments of the trends in biodiversity in an efficient and measurable way so that countries can no longer escape from their responsibilities.[13]

Halting the Loss of Natural Habitats

Preserving the world's forests and natural habitats requires actions at the local, national, and global levels. Unfortunately, these areas are in rapid decline. The global forest area shrank by 3.4 percent (1.4 million square kilometers) from 1990 to 2010—an area roughly the size of Mexico. Deforestation, mainly the conversion of forests to agricultural land, continues at a high rate in many countries. In addition, the extension of built-up areas and transport networks drives the changes in global land use. At the regional level, Africa and South America suffered the largest net losses of forests since 2000, corresponding to annual losses of 0.5 percent in both continents.[14]

Target 5 of the Strategic Plan for Biodiversity states that "by 2020, the rate of loss of all natural habitats, including forests, is at least halved and where feasible brought close to zero." This somewhat imprecise and not very ambitious target should be reinforced to a target of a complete halt of deforestation and loss of natural habitats. This requires that all nations start to deal with the forces behind the use of more and more land for the production of timber, food, animal fodder, and more recently biofuels. It requires that policies and subsidies that drive deforestation are adapted to a zero deforestation economy.

For example, workers in the (illegal) logging industry should be assigned to jobs that help protect forest ecosystems rather than destroying them. Such an approach has been used in other areas—the TAMAR sea turtle program in Brazil hires ex-turtle poachers, for instance, paying them wages to protect rather than exploit the turtle population. The TAMAR effort now serves dozens of coastal communities in northeastern Brazil by providing employment and other public benefits to local residents. A recent UNEP analysis suggests that an investment of just $40 billion a year from 2010 to 2050 in reforestation and payments to landholders to conserve forests could raise the value added in the forest industry by 20 percent.[15]

By 2011 there were at least 160,000 protected areas in the world, covering about 13 percent of the land area (an area the size of Russia). Marine protected areas, however, cover only around 7 percent of coastal waters and 1.4 percent of the oceans. The Strategic Plan for Biodiversity includes the target of protecting 17 percent of terrestrial and inland waters and 10 percent of coastal and marine areas. Yet these targets are far from ambitious, and the plan also lacks a framework for ensuring that a designated area is actually protected.[16]

In theory, protected areas such as reserves and national parks are useful because they allow little to no resource extraction and they minimize or prohibit development. In practice, however, protected areas often tell a different story. In Indonesia, for example, an estimated

12 million hectares of tropical forest are supposedly protected, but the picture on the ground is very different as forests continue to be logged and burned down. Protected area boundaries are proving a poor defense against illegal logging, agricultural encroachment, and poaching. According to a satellite-based analysis, some 1.3 million hectares of low-access forest in Indonesia are simultaneously protected and within logging concessions, clearly illustrating the government's inability to uphold conservation policies.[17]

The situation for the oceans is likewise alarming. A majority of the world's coral reefs are under severe threat from climate change impacts as well as unsustainable fishing. (See Box 15–2.) The world's fishing fleet is thought to be able to catch up to 2.5 times sustainable levels. Industrial fishing with trawls from large-scale vessels is especially damaging to ocean health and species diversity. Bringing fishing yields to sustainable levels requires decisive measures. In the European Union and elsewhere, subsidies to industrial fishing should be phased out or redirected to sustainable practices that will help the environment and bring benefits to local communities.[18]

New global agreements under the U.N. Convention on the Law of the Sea are needed for the conservation of marine biodiversity within sea areas of national jurisdiction but also beyond those, given that today they remain unprotected and unregulated. In the same manner, a global network of marine reserves should be implemented to raise the modest percentage of the oceans protected today. Conserving at least 20 percent of global oceans—including all major hot spots of marine biodiversity, such as coral reefs and seamounts—should be agreed to at the Rio+20 Conference or soon thereafter. The UNEP study on the green economy indicated that greening world fisheries and protecting marine resources better could increase global resource

---

Box 15–2. Coral Reefs Under Threat

Coral reefs are often referred to as the "rainforests of the ocean" due to their immense biodiversity. Different species of coral build structures of various sizes and shapes, creating an exceptional variety and complexity within the coral reef ecosystem and providing habitat and shelter for a large diversity of sea organisms.

Coral reefs are showing increasing signs of stress, however, especially the ones located near coastal developments. About one fifth of the world's coral reefs have already been lost or severely damaged, while another 35 percent could be lost in 10–40 years. Many of the ongoing threats to coral reefs can be linked to human activities, including overfishing and destructive fishing practices. Climate change impacts have been identified as one of the greatest threats to coral reefs. As the temperature rises, mass bleaching and infectious disease outbreaks are likely to become more frequent. In addition, increased levels of atmospheric carbon dioxide alter seawater chemistry by causing acidification. As seawater becomes more acid, organisms with a calcium carbonate (limestone) skeleton, like polyps the building organisms that are the basis of corals—will find it more difficult to maintain their growth. In extreme situations their shell or skeleton may even start to dissolve.

Scientists' understanding of the biological consequences of ocean acidification is at an early stage. So far the only effective way to avoid acidification is to prevent carbon dioxide buildup in the atmosphere through reductions in fossil fuel emissions. Hence, saving coral reefs requires not only better regulation to protect them from unsustainable fishing but also serious attention to the problem of climate change.

—*Eirini Glyki and Bo Normander*
Source: See endnote 18.

rents from negative $26 billion to positive $45 billion a year, hence contributing to increased economic prosperity.[19]

Real Changes Needed

To achieve successful protection of both terrestrial and marine biodiversity, it is absolutely key that already designated as well as newly assigned areas are far better protected and that the local and national authorities are allocated the resources and means to protect the land and sea. It is a political issue for many countries, and something that needs to be fought for at both the global and the national level. But as important as the protection of natural habitats and the implementation of ambitious biodiversity targets are, lowering the unsustainable consumption rate per person, especially in industrial nations, is just as crucial. Society currently measures success by economic growth, and growth is measured by an increase

in consumption. (See Chapter 11.) The current model of consumer societies is destroying the planet and its resources, and this must change in order for the planet to be sustained for future generations.

Combating the sixth mass extinction will require a number of concrete measures, as outlined in this chapter, to protect the world's common biological wealth. It will also require some fundamental changes in the way people consume natural resources. And, finally, it will demand that politicians stand up and start making real decisions that can help protect nature and biodiversity and at the same time be an ignition point for creating sustainable prosperity. The Rio+20 Conference in June 2012 is a great chance for the world's political leaders to come together and take the necessary steps to make the good-intentioned talking about green economy and sustainable development turn into some real measures that can help sustain prosperity and save the planet.

CHAPTER 16

Ecosystem Services for Sustainable Prosperity

Ida Kubiszewski and Robert Costanza

W e live in an age of globalization. An age where information travels instantly around the world. Where humans and their built infrastructure have reached every part of the globe, striving for unending material growth and prosperity. These goals would only be possible, however, within a system unconstrained by any biophysical limits. On Earth, we must live within the planetary boundaries set by the functioning of our ecological life-support system.[1]

In pursuit of unending material growth, western society has increasingly favored institutions that promote the private sector over the public and commons sectors, capital accumulation by the few over asset building by the many, and finance over the production of real goods and services. Steady decline in median income and marginal tax rates have reduced the funds available to spend on public goods while simultaneously contributing to rising income disparity and ecosystem degradation. At the same time, many developing countries are on a path to replicate this system, creating a more extreme version of this disparity within their own boundaries.[2]

This view of what "prosperity" means emerged when the world was still relatively empty of humans and their built infrastructure. Natural resources were abundant, social settlements were sparser, and inadequate access to infrastructure represented the main limit on improvements to human well-being. Much has changed in the last century, however. The human footprint has grown so large that in many cases real progress is constrained more by limits on the availability of natural resources and ecosystem services than by limits on built capital infrastructure.[3]

In a full world, we can no longer focus on valuing certain aspects of society while ignoring others. We need to redefine prosperity to ensure that we are moving in the appropriate direction. We first have to remember that the end goal of an economy is to improve human well-being and the quality of life sustainably. Material consumption and gross domestic product (GDP) are merely means to that end, not ends in themselves. We have to recognize, as both ancient wisdom and new psychological research tell us, that material consumption beyond real need can actually reduce overall well-being. We have to be able to distinguish real poverty in terms of

---

Ida Kubiszewski is a research assistant professor and **Robert Costanza** is Distinguished University Professor of Sustainability at the Institute for Sustainable Solutions at Portland State University.

low quality of life versus merely low monetary income.[4]

But most important, we have to identify what really does contribute to human well-being—namely the ecological systems that provide us with fresh water, soil, clean air, a stable climate, waste treatment, pollination, and dozens of other essential ecosystem services. Ecosystem services can be defined as the ecological characteristics, functions, or processes that directly or indirectly contribute to human well-being—the benefits people derive from functioning ecosystems.[5]

Importance of Natural Capital and Ecosystem Services

The ecosystems that provide these various services are sometimes referred to as "natural capital," using the general definition of capital as a stock that yields a flow of services over time. In order for these benefits to be realized, natural capital must be combined with other forms of capital that require human actions to build and maintain. These include built or manufactured capital, human capital, and social or cultural capital.[6]

So how do we identify and determine the importance of the contributions of natural capital to human well-being in a way that will help society use this knowledge to make decisions? One way is to identify the services that ecosystems provide to humans. Even without any subsequent valuation, just knowing about the existence and benefit to humans of the services derived from an ecosystem can help ensure appropriate recognition of the full range of potential impacts of a given policy option. This can make the analysis of ecological systems more transparent and can help inform decisionmakers about the relative merits of different options before them.

Recognition of their existence is nevertheless not enough if the value of those services is not used in decisionmaking by policymakers or consumers. By not having a number attached to the contributions of these services in terms comparable with economic services and manufactured capital, the value of ecosystem services is often perceived to be zero. Hence, they are often given too little weight in policy decisions and usually a lower priority than economic goods and services.

Valuing Ecosystem Services

Why is it so important to value these services in a comparable way? When it comes to decisionmaking, ecological conflicts arise from two sources: scarcity and restrictions in the amount of ecosystem services that can be provided and distribution of the costs and benefits of the provisioning of the ecosystem services. Ecosystem services science makes trade-offs explicit and thus facilitates management and planning discourse. It helps stakeholders make sound value judgments. Ecosystem services science thus generates relevant socioecological knowledge for stakeholders and decisionmakers along with sets of planning options that can help resolve sociopolitical conflicts.[7]

Accurately valuing ecosystem services is one challenge. Another is that many ecosystem services are public goods. This means that they are non-excludable and that multiple users can simultaneously benefit from using them. Such a characteristic poses a problem, as society does not have the institutions and policies to deal with this type of resource. This creates circumstances where individual choices are not the most appropriate approach to valuation. Instead, some form of community or group choice process is needed.

In recent years, scientists and economists have tried to develop techniques for estimating the benefits from ecosystems. Valuation can be expressed in multiple ways, including monetary units, physical units, or indices. Economists have developed a number of valuation methods that typically use metrics expressed in monetary units, while ecologists and others have developed measures or indices expressed in a variety of non-monetary units, such as biophysical trade-offs.[8]

One of the first studies to estimate the value of ecosystem services globally was published in the journal *Nature* in 1997, entitled "The Value of the World's Ecosystem Services and Natural Capital." The authors estimated the value of 17 ecosystem services for 16 biomes to be in the range of $16–54 trillion per year, with an average of $33 trillion per year—a figure larger than annual global GDP at the time.[9]

More recently the concept of ecosystem services gained attention with a broader academic audience and the public when the Millennium Ecosystem Assessment (MA) was published in 2005. The MA was a four-year study that involved 1,360 scientists and that was commissioned by the United Nations. The report analyzed the state of the world's ecosystems and provided recommendations for policymakers. It determined that human actions have depleted the world's natural capital to the point that the ability of a majority of the "planet's ecosystems to sustain future generations can no longer be taken for granted."[10]

In 2008, a second international study was published on The Economics of Ecosystems and Biodiversity (TEEB), hosted by the United Nations Environment Programme. TEEB's primary purpose was to draw attention to the global economic benefits of biodiversity, to highlight the growing costs of biodiversity loss and ecosystem degradation, and to draw together expertise from the fields of science, economics, and policy to enable practical actions moving forward. The TEEB report was picked up extensively by the mass media, bringing ecosystem services to a broad audience.[11]

Even though much new research and

Kevin Muckenthaler

Carbon storage: maple trees in Olympic National Park's Hoh Rain Forest, Washington State

reporting is being done around the topic of ecosystem services, uncertainty always exists in measurement, monitoring, modeling, valuation, and management. To reduce this, constant evaluation is necessary to determine the impacts of existing systems and to design new systems with stakeholder participation as experiments from which we can more effectively quantify performance and learn ways to manage such complex systems.

A key challenge in any valuation is imperfect information. Individuals might, for example, place no value on an ecosystem service if they do not know the role that the service plays in their well-being. Here is an analogy. If a tree falls in the forest and there is no one around to hear it, does it make a sound? The answer to this age-old question obviously depends on how "sound" is defined. If sound is the perception of sound waves by people, then the answer is no. If sound is defined as the pattern of physical energy in the air, the answer is yes. In the case of ecosystem services, individuals' actions and stated preferences would not reflect the true benefit of ecosystem services as they do not realize the existence of the benefits being provided. Another important challenge is accurately measuring the functioning of a system to correctly quantify the amount of a given service derived from that system.[12]

But recognizing the importance of ecosystem services does not eliminate the limitations that human perception–centered valuation creates. As the tree analogy demonstrates, perceived value can be a quite limiting valuation criterion, because natural capital can provide positive contributions to human well-being that are either never (or only vaguely) perceived or may only manifest themselves in the future. A broader notion of value allows a more comprehensive view of value and benefits, including, for example, valuation relative to alternative goals/ends, such as fairness and sustainability, within the broader goal of human well-being. Whether these values are perceived or not and how well or accurately they can be measured are separate and important questions.[13]

Ecosystem service: drawing water from the Ogallala aquifer, Buffalo Lake National Wildlife Refuge, Texas

The incorporation of the value of ecosystem services into the definition of sustainable prosperity is critical to ensuring that a "real" and sustainable prosperity can be estimated and

pursued. It goes beyond that, however: ecosystem services are essential to the existence of human society, as they are the life support system of the planet. Often the connection between ecosystem services and human health, and hence prosperity, is difficult to make, since it can be indirect, displaced in space and time, and dependent on many forces.[14]

Institutions around Ecosystem Services

Recognizing that we are in a biophysical crisis because of our overconsumption and lack of protection of ecosystem services, we must invest in institutions and technologies to reduce the impact of the market economy and to preserve and protect public goods. New types of institutions are needed to do this, using a sophisticated suite of property rights regimes. We need institutions that use an appropriate combination of private, state, and common property rights systems to establish clear property rights over ecosystems without privatizing them.

One such category of institution is the commons sector, which would be responsible for managing existing common assets and for creating new ones. Some assets should be held in common because it is more just; these include resources created by nature or by society as a whole—for example, a freshwater environment created by nature or common knowledge created by society. Others should be held in common because it is more efficient; these include nonrival resources for which price rationing creates artificial shortages (information) or rival resources (goods that are used up through consumption) that generate nonrival benefits, such as trees filtering water to make it drinkable. Others should be held in common because it is more sustainable; these include essential common pool resources and public goods such as clean air.[15]

An example of such an institution for managing the commons sector is a "common asset trust" at various scales. Trusts can "propertize" the commons without privatizing them, as do many land trusts currently in existence. Common asset trusts could protect and restore critical natural capital—the resources provided by nature that are in some way essential to human well-being. They can also promote information and technologies that can protect or enhance public goods. Examples of this include low-pollution energy sources, non-ozone-depleting refrigerants, organic agriculture, erosion- and drought-resistant agriculture (such as perennial grains), alternatives to trawl fishing, devices that reduce bycatch in fisheries, and so on. All such information should be freely available for whoever chooses to use it.[16]

Another such institution that has provided a model of this type of institution is "payment for ecosystem services." This sets up a system in which landowners or farmers are paid to maintain the ecosystems that provide services to the rest of the population in a region. Those using the services provide the money for the payment. Probably the best known such system was implemented in Costa Rica over a decade ago: landowners are paid to plant or preserve forested areas on their land. A workshop was held in Costa Rica around this issue and proved to be very successful.[17]

Ideas about ecosystem services and their valuation have begun to appear not only in public media outlets in the form of high-profile reports but also in the business community. Dow Chemical recently established a $10-million collaboration with The Nature Conservancy to tally up the ecosystem costs and benefits of every business decision. Such collaboration will provide a significant addition to ecosystem services valuation knowledge and techniques. But significant research and new institutional design are still required.[18]

Hundreds of projects and groups are currently working toward better understanding, modeling, valuation, and management of ecosystem services and natural capital. It would be impossible to list all of them here, but a few key ones are a new international Ecosystem Services Partnership that is a global network helping to coordinate activities and build consensus; a World Bank initiative called Wealth Accounting and Valuation of Ecosystem Services, with the goal of improving information available to decisionmakers in Ministries of Finance and Planning or in central banks so that development can proceed in a more sustainable fashion; and a new United Nations effort called the Intergovernmental Platform on Biodiversity and Ecosystem Services that will be an interface between the scientific community and policymakers and that aims to build capacity for and strengthen the use of science in policymaking.[19]

Priorities on Ecosystem Services

Given that significant levels of uncertainty exist in ecosystem service measurement, monitoring, modeling, valuation, and management, we should continuously gather and integrate appropriate information, with the goal of learning and adaptive improvement. To do this we should constantly evaluate the impacts of existing institutions and design new ones with stakeholder participation as experiments from which we can more effectively quantify performance and learn.

We need institutions that can effectively deal with the public goods nature of most ecosystem services, using a more sophisticated suite of property rights regimes. We need institutions that use a balanced combination of existing private property rights systems and new systems that can propertize ecosystems and their services without privatizing them. Systems of payment for ecosystem services and common asset trusts can be effective elements in these institutions.

The spatial and temporal scale of the institutions to manage ecosystem services must be matched with the scales of the services themselves. Mutually reinforcing institutions at local, regional, and global scales over short, medium, and long time scales will be required. Institutions should be designed to ensure the flow of information between scales, to take ownership regimes, cultures, and actors into account, and to fully internalize costs and benefits.

Distribution systems should be designed to ensure inclusion of the poor, since they depend more on common property assets like ecosystem services. Free-riding should be prevented, and beneficiaries should pay for the services they receive from biodiverse and productive ecosystems.

One key limiting factor in sustaining natural capital is shared knowledge of how ecosystems function and how they support human well-being. This can be overcome with targeted educational campaigns, clear dissemi-

nation of success and failures directed at both the general public and elected officials, and true collaboration among public, private, and government entities.

Relevant stakeholders—local, regional, national, and global—should be engaged in the formulation and implementation of management decisions. Full stakeholder awareness and participation contributes to credible, accepted rules that identify and assign the corresponding responsibilities appropriately and that can be effectively enforced.

Ecosystem concepts can be an effective link between science and policy by making the trade-offs in today's world more transparent. An ecosystem framework can therefore be a beneficial addition to policymaking institutions and frameworks and can help to integrate science and policy.

These are just first steps. But in order to establish sustainable prosperity for all, the value of ecosystems services will need to be understood and factored into all policy and business decisions in the future.

CHAPTER 17

Getting Local Government Right

Joseph Foti

Nchunu Justice Sama is a barrister practicing in Bamenda, Cameroon. Starting in 2005, he watched as the dump at Atuanki, Mile 6 Mankon, grew at the edge of the Mezam River. Untreated waste from the site leached into nearby settlements, leaked into the river, and spread across the adjacent highway. Sama and his associates wrote to the City Council and to the neighborhood councils, using the formal petition system to ask for enforcement of solid waste laws. They received no reply.[1]

The silence spurred Sama's organization, the Foundation for Development and Environment (FEDEV), to use the local courts. Together with a team of public interest lawyers, Sama demanded that the local governments stop dumping near the settlement or the river, begin cleanup to relocate the waste to a dump operating within legal requirements, and provide information to the public about solid waste disposal. As they began the process of litigation, each of the local city councils argued that the dump was in another jurisdiction. In fact, it was unclear: members of the public did not have adequate maps of jurisdiction for the local councils. It became necessary to bring all the local councils to court.

While FEDEV was not the first Cameroonian organization to initiate litigation in the public interest, no organization or citizen in the country had successfully used the courts to enforce environmental laws where they could not prove personal harm. When the case of *FEDEV & 1 Other v. Bamenda City Council & 2 Others* reached the High Court, it tested this precedent. After several months of litigation, the final ruling from the courts was announced:

> [D]umping waste on the surface land at Atuanki, mile 6 Mankon, and the pollution of Mezam River are infringements of fundamental rights of citizens. Without deciding on the merits or demerits of the case, I agree with Learned Counsel for the Plaintiff that the protection of fundamental human rights is the exclusive preserve of ordinary law courts. I also agree with them that any act of degradation, of environment, by whomsoever is an infringement of the rights of citizens to a healthy environment.... From the foregoing I hold that this court has the competence to hear and determine the questions posed in the

Joseph Foti is a senior associate at the World Resources Institute in Washington, DC, which serves as the global secretariat for the Access Initiative.

Both, Nchunu Justice Sama

Left: *The illegal dump at Atuanki, Bamenda, Cameroon.* Right: *The cleaned up site following the High Court ruling in* FEDEV v. Bamenda City Council.

plaintiff's origination summons and to grant the relief sought.[2]

The ruling was groundbreaking in two respects. It was the first time a Cameroonian High Court had ruled that the right to environment was a fundamental human right. And, perhaps more important, it allowed any member of the public to use the courts to enforce an environmental law. In the months that followed, the illegal dump was cleaned and a newer, safer dump was constructed.

The case of *FEDEV v. Bamenda City Council* is emblematic of the kind of little victories that must be hard-won thousands of times over, across the world, as each country works to put in place measures to move toward sustainable development. There is no magic bullet to clean up the environment or improve the lives of citizens. Instead, there are many little decisions to be made. Governments cannot do it alone. Citizens need to ensure that laws are enforced and that environment and equity are considered when making development decisions.

When 172 governments met at the United Nations Conference on Environment and Development in Rio de Janeiro in 1992, they agreed that:

Environmental issues are best handled with the participation of all concerned citizens, at the relevant level. At the national level, each individual shall have appropriate *access to information* concerning the environment that is held by public authorities, including information on hazardous materials and activities in their communities, and the *opportunity to participate* in decision-making processes. States shall facilitate and encourage public awareness and participation by making information widely available. Effective *access to judicial and administrative proceedings*, including redress and remedy, shall be provided [emphasis added].[3]

This section of the Rio Declaration, Principle 10, is sometimes referred to as the Environmental Democracy Principle. It contains some of the basic building blocks of modern environmental management. (See Box 17–1.) Since the first Rio Conference, national governments have made huge strides in implementing Principle 10—from making public participation in project planning nearly universal to the passage of the Convention on Access to Information, Public Participation in Decision-making and Access to Justice in Environmental Matters (known as the Aarhus Convention), the only legally binding treaty on environmental democracy.[4]

Not all decisionmaking for sustainable development takes place at the national level. Local

Box 17–1. The Elements of Principle 10 at the Local Level

Access to Information. Refers to the availability of information on the environment and the mechanisms by which public authorities provide environmental information. At the local level, examples of access to information include regular information on air and water quality, local decisionmaking, land use, and permitting data.

Public Participation. Refers to the opportunities for individuals, groups, and organizations to provide input to decisionmaking that will have—or is likely to have—an impact on the environment. At the local level, public participation can be integrated into policymaking, land use planning, permitting, and project-level decisions.

Access to Justice. Refers to effective judicial and administrative procedures and remedies available to individuals, groups, and organizations for actions that affect the environment and contravene laws or rights. The legal standing to sue and the ability to litigate are components of access to justice. At the local level, access to justice means that local authorities have impartial, inexpensive, and efficient institutions that hear complaints of denial of information, environmental harm, and noncompliance with the law. These may be courts or administrative tribunals or petition systems.

Source: See endnote 4.

Agenda 21s, the blueprint for implementing sustainable development at the local level, recognize the vital importance of local authorities. They "construct, operate and maintain economic, social and environmental infrastructure, oversee planning processes, establish local environmental policies and regulations, and assist in implementing national and sub-national environmental policies. As the level of governance closest to the people, they play a vital role in educating, mobilizing and responding to the public to promote sustainable development." At their best, decisions at the local level promise poverty reduction, job growth, gender equity, and environmental improvement—each a critical component of sustainable development.[5]

The Challenge of Local Democracy

Local democracy, especially in cities, is critical to sustainable development. Cities in particular will be engines of sustainable development, critical to growing in a less resource-intensive manner while reducing poverty. (See also Chapters 3 and 5.) Between 2000 and 2050, growth of the urban population will outpace total population growth, meaning the shared future is significantly more urban. Most of the new growth will be in the cities of the developing world. This urbanization is strongly associated with poverty reduction.[6]

But at the same time as cities grow and poverty is reduced, cities also may experience more environmental impact (due to increased consumption), inequality, and insecurity. To manage the growth process in a fair manner and to ensure that environment and poverty reduction remain central to growth, decisions must be open to voices advocating for these things. In the multiethnic, globally competing, often-segregated cities of today and tomorrow, the way in which decisions are made can be as important as the final decisions.

The pressures of developing cities sustainably can best be managed when local institutions—especially government authorities—are transparent, participatory, and accountable. Such institutions are more efficient, limiting undue influence and corruption. Participation also encourages sustainable development. Stakeholders can present solutions that were previously not considered and authorities can allocate resources in a manner that better reflects public demand. Such decisions are more legitimate and, as a consequence, more robust. There is strong evidence to suggest that where people feel that a fair process is in place, they are more willing to accept decisions they disagree with. Local democracy promotes sustainability at other scales as well, as it is often the crucible of democracy and policy innovation at a larger scale.[7]

Despite international consensus on the importance of transparency, participation, and accountability at the local level, progress has been uneven. While some local governments have been innovators who have implemented Principle 10, others have lagged. This creates a significant barrier to sustainable development in many cases, as key decisions about land use (zoning and location of polluting industries), provision of safe drinking water, waste management, and resource extraction (such as mine permits and contracts) are often devolved to the local level.[8]

In many places, local institutions are weak or unaccountable. Often, decentralization—the process of shifting decisionmaking from the national capital to levels of government closer to the people—remains an incomplete process, and local governments might not have the power to pass new laws or the resources to carry out their jobs. Where decentralization has occurred, otherwise democratically accountable local authorities may be crowded out of their mandated role by nondemocratic systems: traditional decisionmaking systems, informal systems, other governments, nongovernmental organizations, or the private sector—none of which are subject to the same controls for public accountability. In other cases, local institutions are not democratic, either by law or in practice (perhaps as a result of tampered elections).

The Opportunity of Rio+20

Twenty years after Principle 10 was first agreed to, the governments of the world will meet again at Rio de Janeiro for Rio+20. This con-ference offers the promise of renewed commitment and collaboration around the principles for sustainable development. One of the key themes will be the Institutional Framework for Sustainable Development, where governments will decide on the shape and practices of governing for more equitable and inclusive development. While a great deal of focus has been placed on the international and national levels, there is growing interest in improving governance and decisionmaking at local levels as well. Rio+20 can serve as a platform for innovative commitments to better governance by local authorities.

Civil society groups around the world concerned with better local governance for sustainable development have worked together to formulate a proposal for Rio+20. Many of the organizations behind this proposal are members and affiliates of the Access Initiative. (See Box 17–2.) They seek to accelerate implementation of Principle 10 at all levels, including the local.[9]

To diagnose the specific challenges of making local authorities more transparent, participatory, and accountable, the group gathered case studies highlighting local urban struggles from Argentina, Bolivia, Cameroon, Chile, Costa Rica, Ecuador, Hungary, Mexico, Thailand, and the United States to identify common barriers to the ability to participate in sustainable development decisions. The organization also identified innovative approaches to bringing local communities and sustainability advocates into the decisionmaking process, such as participatory budgeting, social audits, and citizen suits already in place in many local jurisdictions.

While the cases examined cannot make claims to representativeness, they can provide some idea of the major barriers to urban sustainability as well as the innovations. Table 17–1 shows a number of highly successful instances of protection of the environment, preservation of heritage, and defense of the interests of poor or disadvantaged communi-

Box 17–2. The Access Initiative

The Access Initiative is the world's largest network of civil society organizations dedicated to ensuring that local communities have the rights and abilities to gain access to information and to participate in decisions that affect their lives and their environment. Members from around the world carry out evidence-based advocacy to encourage collaboration and innovation that advances transparency, accountability, and inclusiveness in decisionmaking at all levels. On issues from freedom of information laws to participation in environmental impact assessment, from ensuring that isolated communities have the ability to affect policy decisions to opening courts to serve the public in cases of environmental harm, the organizations that belong to the Access Initiative work to tie local struggles to reform at all levels, helping build environmental democracy.

Source: See endnote 9.

ties. In each case, institutions and rules were in place that allowed advocates to move society to a more sustainable path.[10]

Recent experience with government transparency in the United States illustrates how opening a new data set (in this case, on federal spending) can aid more-sustainable transport and environmental cleanup spending. The federal government responded to the economic crisis with the American Recovery and Reinvestment Act, or "stimulus bill," aimed at creating spending to improve job growth. Citizens across the country became concerned with the speed and efficiency of those expenditures, as well as the distribution of funds to various programs. Much of the U.S. federal budget is granted to local authorities to implement projects, but it has historically been difficult to track timelines and expenditures for individual projects.

To encourage public policy advocacy in addressing transportation equity, the advocacy organization OMB Watch developed the Equity and Government Accountability Project (EGAP), a Web-based application that combines census data with data available from federal websites such as FedSpending.org, including Recovery Act transportation investments. Data are available on an interactive map at the level of states, counties, or congressional districts. This has allowed people to find out where government money has gone and compare that with their community needs. In addition, they will be able to explore how effective these programs have been in meeting the needs of specific communities.[11]

In Missouri, for example, community activists monitored the implementation of the $500 million I-64 highway project. With this information, they were able to publicize ongoing proceeds and to attend public events. As a consequence, the project came in $11 million under budget, employed 26 percent minority and female workers, and contained the largest community benefits agreement in U.S. history. This success has led to adoption of similar agreements in other cities and similar advocacy around transit. With respect to environmental cleanup, EGAP helped community organizations from disadvantaged communities track more than $600 million in funding and spending on hazardous waste cleanup in the most economically depressed areas, where environmental health impacts are often most acute. This project demonstrates the impact that the powerful combination of transparent government and an active civil society can have on sustainable development, job growth, and poverty reduction.[12]

Where Do We Go from Here?

The case studies demonstrated a number of barriers to transparency, accountability, and inclusiveness at the local level. In Cameroon,

Table 17–1. Summary of Case Studies in Urban Governance

Case	Issue Area	Innovations
Access to Information		
Sanitation in Población Gabriel Gonzalez Videla (Santiago, Chile)	Waste management	Public access to information on sewage and sanitation
Saving the Buda castle (Budapest, Hungary)	Historic preservation	Disclosure of construction permits
Construction in conservation zone (Mexico City, Mexico)	Land use	Release of construction and conservation plans
The Equity and Government Accountability Project (United States)	Spending and transit	Access to government spending and demographic data
Public Participation		
Riachuelo-Matanza River Basin Management (Buenos Aires, Argentina)	Water quality	Public participation in integrated water resources management
Illegal settlements and landslides (La Paz, Bolivia)	Housing and land use	Public participation in land use planning and resettlement
Coastal zone planning and tourism (Tarcoles, Costa Rica)	Coastal zone planning	Public participation in coastal zone planning
Conversion of planned supermarket to park through impact assessment (Cuernavaca, Mexico)	Land use	Public participation in environmental impact assessment
Bypass suspended (Cuernavaca, Mexico)	Transportation and land use	Public participation in environmental impact assessment
Urban transport (Guadalajara, Mexico)	Transit	Public participation in strategic transit planning
Multistakeholder panel for pollution control (Map Tha Phut City, Thailand)	Air and water pollution	Public participation in pollution reduction planning
Access to Justice		
Waste management: City Council (Bamenda, Cameroon)	Waste management	Extension of public interest standing to civil society organizations
Writ of Amparo for the Ayora Community fighting a landfill (Ecuador)	Waste management	Right to environment invoked through courts
Stopping a shopping mall (Budapest, Hungary)	Land use	Broad standing for public interest litigation
Freeway war on the Danube (Budapest, Hungary)	Transportation	Broad standing for public interest litigation
Case of Lerma Tres Marias: Injunction on highway construction (Texcalyacac, Mexico)	Transportation and land use	Use of injunctive relief to suspend project
Environmental Procurator General suspends works on a public project (Mexico City, Mexico)	Transportation and land use	Ombudsman or public officer to enforce environmental laws
The legal case of highway construction (Fierro del Toro, Mexico)	Transportation and land use	Court admits a "coadvuyancia" (similar to an amicus brief)

Source: See endnote 10.

for example, the public was unable to identify the agency responsible for the issue at hand. In Chile and the United States, the promised services were not delivered and the public could not trace how allocated funds had been spent. In Thailand, decisions were made in secret, or the public was brought into a decisionmaking process well after all decisions had already been made. In the land use and transportation cases in Mexico, decisions were made at the national level; local authorities and local residents had little say in their design and implementation. And in Argentina the public lacked the data they needed in order to participate in complicated decisions such as river basin management.[13]

Based on this analysis, governments can take a number of concrete steps to surmount these barriers and advance transparency, inclusiveness, and accountability at the local level:

- *Access to Information:* Make information on all agency jurisdictions, budgeting, revenue, and procurement available and usable; adopt local Access to Information laws, providing a mechanism for request of government-held information; pass open meeting laws for all local authorities; provide proactive information on land use, development planning, transportation, waste disposal, utilities, and regular environmental quality monitoring data.

- *Public Participation:* Accept and promote mechanisms for public accountability in service delivery, such as public social audits and report cards of agency performance; adopt reforms for early, meaningful public participation in policy and planning by a broad range of stakeholders; expand the number of decisions that incorporate public participation and oversight; build the capacity of stakeholders to participate by integrating the rights

and means to obtain information and participation into educational curricula.

- *Access to Justice:* Improve the authority and capacity of ombudspersons and other law enforcement officers to monitor and enforce environmental laws and civil rights protections; create public interest standing and mechanisms to allow citizen enforcement of environmental laws.

Rio+20 provides a platform for public officials at all levels to commit to innovative implementation of Principle 10. Mayors and local executives attending the conference can commit publicly to the reforms just described. National governments can commit to reforms that encourage decentralization to democratic local institutions. They can also accelerate implementation of Principle 10 at the local level by creating an encouraging legal and administrative environment, supporting innovation and building the capacity of officials. Finally, Rio+20 may provide an opportunity to expand international legal mechanisms like the Aarhus Convention to improve accountability at all levels of government.

Regardless of the results of Rio+20, there is widespread need to improve decisionmaking processes at the local level. Mitigating and responding to the threats of global climate change, water scarcity, and dwindling natural resources requires governments and civil society to manage inevitable trade-offs at all levels, but especially locally. To move toward environmental sustainability, people need strong institutions and an ability to work collaboratively across sectors. To ensure that environmental sustainability is actually politically sustainable and both economically and socially just, decisions must be reached through transparent, democratic means.

Notes

State of the World: A Year in Review

October 2010. Juliette Jowit, "Western Lifestyles Plundering Tropics at Record Rate, WWF Report Shows," (London) *Guardian*, 13 October 2010; Neela Banerjee, "GM to Set Aside $773 Million to Clena Up Old Factory Sites," *Los Angeles Times*, 21 October 2010; "Elephant Declared National Heritage Animal," *Press Trust of India*, 22 October 2010; "Nations Agree on Historic UN Pact on Sharing Benefits of World's Genetic Resources," *UN News Centre*, 29 October 2010.

November 2010. "Voters Reject Prop. 23, Keeping California's Global Warming Law Intact," *Los Angeles Times–PolitiCal*, 2 November 2010; Olivia U. Mason et al., "First Investigation of the Microbiology of the Deepest Layer of the Ocean Crust," *PLoS One*, 5 November 2010; Earthworks, "Indigenous Peoples in Latin America Call for an End to Destructive Mining," press release (Lima, Peru: 26 November 2010); "Latin America Suffers Intense and Deadly Rains," *Environment News Service*, 30 November 2010.

December 2010. John Vidal, "Cancún Climate Change Summit: Japan Refuses to Extend Kyoto Protocol," (London) *Guardian*, 1 December 2010; "Wildfire in Israel," The Big Picture, *Boston.com*, at www.boston.com/bigpicture/2010/12/wildfire_in _israel.html; EarthRights International and Amazon Watch, "Indigenous Peruvians Win Appeal in Federal Human Rights and Environmental Lawsuit Against Occidental Petroleum for Contaminating Amazon Rainforest, Poisoning Communities," press release (Los Angeles: 6 December 2010); Mary Esch, "NY 'Fracking' Ban: Governor David Paterson Orders Natural Gas Hydraulic Frackuring Moratorium for Seven Months in New York," *Huffington Post*, 12 December 2010; "US Senate Approves Nuclear Arms Control Treaty with Russia," (London) *Guardian*, 22 December 2010.

January 2011. "Scientists Find 'Drastic' Weather-related Atlantic Shifts," *Terra Daily*, 3 January 2011; Maggie Fox, "Researchers Find 'Alarming' Decline in Bumblebees," *Planet Ark*, 4 January 2011; "2010 Tied for Earth's Warmest Year on Record," *Environment News Service*, 12 January 2011; Jennifer LaRue Huget, "Nearly All Pregnant Women Harbor Potentially Harmful Chemicals" (blog), *Washington Post*, 14 January 2011; "Australia Makes Green Cuts to Fund Flood Relief," *Radio Australia*, 28 January 2011.

February 2011. Tan Ee Lyn, "Growing Number of Farm Animals Spawn New Diseases," *Reuters*, 11 February 2011; "UNEP: $1.3 Trillion a Year Would Turn World Economy Green," *Environment News Service*, 21 February 2011; "Rare, Unique Seeds Arrive at Svalbard Vault, as Crises Threaten World Crop Collections," *EurekAlert*, 25 February 2011; IndiaResources.org, "Kerala Passes Law Allowing Compensation from Coco-Cola," *CounterCurrents.org*, 24 February 2011.

March 2011. Lyndsey Layton, "Scientists Want to Help Regulators Decide Safety of Chemicals," *Washington Post*, 4 March 2011; Alan Taylor, "Japan Earthquake: Six Months Later," *The Atlantic*, 12 September 2011; Luke O'Brien, "Green Sports Alliance: Go Green or Go Home," *Fast Company*, September 2011; Will Nichols, "EU to Drive Petrol Cars from Cities by 2050," *Business Green*, 28 March 2011; Ruth Dasso Marlaire, "NASA Satellites Detect Massive Drought

Impact on Amazon Forests," *NASA Feature*, 29 March 2011.

April 2011. "Arctic Ozone Hole Largest in History," *Environment News Service*, 7 April 2011; "Staph Seen in Nearly Half of U.S. Meat," *CNN.com*, 15 April 2011; Geoff Olson, "Bolivia's Law of Mother Earth," *Common Ground*, July 2011; Chuck Squatriglia, "Discovery Could Make Fuel Cells Much Cheaper," *Wired*, 22 April 2011.

May 2011. "Green Party Wins First Seat in Canada's Parliament," *Environment News Service*, 2 May 2011; Brian Merchant, "WikiLeaks: Rush to Drill in Arctic Is Stirring Military Tension with Russia," *Treehugger*, 12 May 2011; "Prince of Wales: Ignoring Climate Change Could Be Catastrophic," (London) *Telegraph*, 24 May 2011; International Energy Agency, "Prospect of Limiting the Global Increase in Temperature to 2ºC Is Getting Bleaker," press release (Paris: 30 May 2011); Juergen Baetz, "Germany Nuclear Power Plants to be Entirely Shut Down by 2022," *Huffington Post*, 30 May 2011.

June 2011. European Commission, "Bisphenol A: EU Ban on Baby Bottles to Enter into Force Tomorrow," press release (Brussels: 31 May 2011); "Autism Experts Urge Reform of U.S. Chemicals Law," *Environment News Service*, 8 June 2011; "Mozambique's Lake Niassa Declared Reserve and Ramsar Site," *PR Web*, 9 June 2011; Leslie Guevarra, "First Take: Google's $280M Solar Fund, GM Invests in Green Buses, and More," *GreenBiz.com*, 14 June 2011.

July 2011. Douglas Fischer, "Economists Find Flaws in Federal Estimate of Climate Damage," *The Daily Climate*, 13 July 2011; "Cosmetic Pesticides Banned in N.L.," *CBC News*, 14 July 2011; Sudarsan Raghavan, "U.N.: Famine in Somalia Is Killing Tens of Thousands," *Washington Post World*, 20 July 2011; Aya Takada, "Japan's Food-Chain Threat Multiplies as Fukushima Radiation Spreads," *Bloomberg.com*, 25 July 2011.

August 2011. Michael Hirtzer, "Drought Worsens in Midwest; Parched Plains in Bad Shape," *Reuters*, 4 August 2011; Eric A. Taub, "Philips Wins Energy Department's Lighting Prize" (blog),

New York Times, 3 August 2011; John Landers, "China Implements a National Feed-in-Tariff Rate," *EnergyTrend*, 12 August 2011, and Coco Liu, "China Uses Feed-in Tariff to Build Domestic Solar Market," *New York Times*, 14 September 2011; "Britain: Oil Leak Stopped in North Sea," *New York Times*, 19 August 2011; Andrew E. Kramer, "Exxon Reaches Arctic Oil Deal with Russians," *New York Times*, 30 August 2011.

September 2011. National Center for Atmospheric Research, "Switching from Coal to Natural Gas Would Do Little for Global Climate, Study Indicates," press release (Boulder, CO: 8 September 2011); Juliann E. Aukema et al., "Economic Impacts of Non-Native Forest Insects in the Continental United States," *PLoS ONE*, 9 September 2011; Annie Snider, "Army Initiative Could Be Boon for U.S. Solar Companies," *New York Times*, 16 September 2011; Thomas Fuller, "Myanmar Backs Down, Suspending Dam Project," *New York Times*, 30 September 2011.

October 2011. Suzanne Goldenberg, "US Must Stop Promoting Biofuels to Tackle World Hunger, Says Thinktank," (London) *Guardian*, 11 October 2011; Rachel Nuwer, "Climate Change Is Shrinking Species, Research Suggests" (blog), *New York Times*, 16 October 2011; Cornelia Dean and Rachel Nuwer, "Salmon-Killing Virus Seen for First Time in the Wild on the Pacific Coast," *New York Times*, 17 October 2011; "Investors Worth $20 Trillion Urge Legally-Binding Climate Treaty," *Environment News Service*, 19 October 2011; "A Child Is Born and World Population Hits 7 Billion," *MSNBC.com*, 31 October 2011.

November 2011. Jia Lynn Yang, "Does Government Regulation Really Kill Jobs? Economists Say Overall Effect Minimal," *Washington Post*, 13 November 2011; Hanna Gersmann and Jessica Aldred, "Medicinal Tree Used in Chemotherapy Drug Faces Extinction," (London) *Guardian*, 9 November 2011; Elizabeth Grossman, "Northwest Oyster Die-offs Show Ocean Acidification Has Arrived," *Yale Environment 360*, 21 November 2011; "Jacob Zuma Opens Durban Climate Negotiations with Plea to Delegates," (London) *Guardian*, 28 November 2011.

Chapter 1. Making the Green Economy Work for Everybody

1.   Tom Bigg, "Development Governance and the Green Economy: A Matter of Life and Death?" in Henrik Selin and Adil Najam, *Beyond Rio+20: Governance for a Green Economy* (Boston: Frederick S. Pardee Center for the Study of the Longer-Range Future, Boston University, 2011), p. 28.

2.   Erik Assadourian, "The Rise and Fall of Consumer Cultures," in Worldwatch Institute, *State of the World 2010* (New York: W. W. Norton & Company, 2010), pp. 3–20.

3.   For the International Labour Organization (ILO) position on the relationship between the financial industry and the rest of the economy, see ILO, *World of Work Report 2011: Making Markets Work for Jobs, Summary*, preprint edition (Geneva: 31 October 2011), p. 2.

4.   Esther Addley, "Occupy Movement: From Local Action to a Global Howl of Protest," (London) *Guardian*, 17 October 2011; "Occupy Together," at www.meetup.com/occupytogether.

5.   Quote is from "About," Occupy COP17 website, at www.occupycop17.org/about.

6.   U.N. Environment Programme (UNEP), *Towards a Green Economy: Pathways to Sustainable Development and Poverty Eradication* (Nairobi: 2011), p. 20.

7.   Ibid.

8.   Johan Rockström et al., "A Safe Operating Space for Humanity," *Nature*, 24 September 2009, pp. 472–75.

9.   Mike Foster, "Economic Growth Puts Global Resources under Pressure," *Financial News*, 3 March 2008.

10. Organisation for Economic Co-operation and Development (OECD), *Perspectives on Global Development 2010: Shifting Wealth* (Paris: 2010), pp. 2, 6; Martin Ravallion, *A Comparative Perspective on Poverty Reduction in Brazil, China and India*, Pol-icy Research Working Paper 5080 (Washington, DC: World Bank, 2009).

11. OECD, op. cit. note 10, p. 6; Brazil comparison from Ravallion, op. cit. note 10, and from Reed M. Kurtz, "Brazil Faces Its Post-Lula Future," North American Congress on Latin America, 1 March 2010.

12. Unemployment and jobless growth from ILO, *Global Employment Trends 2011* (Geneva: 2011), pp. 6, 63; carbon emissions growth from Justin Gillis, "Carbon Emissions Show Biggest Jump Ever Recorded," *New York Times*, 4 December 2011.

13. ILO, *Global Employment Trends January 2010* (Geneva: 2010), pp. 18–19; informal sector earnings from Marc Bacchetta, Ekkehard Ernst, and Juana P. Bustamante, *Globalization and Informal Jobs in Developing Countries* (Geneva: ILO and World Trade Organization, 2009).

14. U.S. wage stagnation and inequality from Economic Policy Institute, Datazone, at www.epi .org/page/-/datazone2008/wage-comp-trends/ earnings.xls; U.S. productivity and wages from Bill Marsh, "The Great Regression: 1980–Now," *New York Times*, 4 September 2011; U. S. poverty from Sabrina Tavernise, "Poverty Levels in 2010 Reach 52-Year Peak, U.S. Says," *New York Times*, 13 September 2011; Germany from Thorsten Kalina and Claudia Weinkopf, "Niedriglohnbeschäftigung 2008: Stagnation auf hohem Niveau–Lohn spektrum franst nach unten aus," IAQ Bericht 2010-06 (Duisburg, Germany: Institut Arbeit und Qualifikation der Universität Duisburg-Essen, 2010), pp. 5, 7; Japan from Machiko Osawa and Jeff Kingston, "Japan Has to Address the 'Precariat,'" *Financial Times*, 1 July 2010; Julia Obinger, "Working on the Margins: Japan's Precariat and Working Poor," Discussion Paper 1, *Electronic Journal of Contemporary Japanese Studies*, 25 February 2009.

15. ILO, op. cit. note 3, p. 3.

16. Figure 1–1 and text from James B. Davies et al., *The World Distribution of Household Wealth*, Discussion Paper No. 2008/03 (Helsinki: UNU-WIDER, 2008), pp. 7–8.

17. Ibid.

18. Germany from Wolfgang Lieb, "Privater Reichtum – öffentliche Armut," *NachDenkSeiten*, 23 June 2010; India from "Key Facts: India Rising," *BBC News Online*, 22 January 2007; Edward N. Wolff, *Recent Trends in Household Wealth in the United States: Rising Debt and the Middle-Class Squeeze—An Update to 2007*, Working Paper No. 589 (Annandale-on-Hudson, NY: Levy Economics Institute of Bard College, 2010), pp. 11, 33.

19. Nick Robins, Robert Clover, and Charanjit Singh, *A Climate for Recovery* (London: HSBC Global Research, 2009), and corrected summary table "The Green Dimension to Economic Stimulus Plans," 26 February 2009.

20. New Economics Foundation, *A Green New Deal* (London: 2008); UNEP, *Global Green New Deal, Policy Brief* (Geneva: 2009); UNEP, *Green Jobs: Towards Decent Work in a Sustainable, Low-Carbon World* (Nairobi: 2008); UNEP, op. cit. note 6.

21. UNEP, op. cit. note 6, pp. 15–16.

22. Box 1–1 from the following: decoupling measurement from UNEP, *Decoupling Natural Resource Use and Environmental Impacts from Economic Growth* (Nairobi: 2011), and from UNEP, op. cit. note 6; global energy intensity from World Bank, GDP (2000 dollars), at data.worldbank.org/indicator/NY.GDP.MKTP.KD, and from British Petroleum (BP), *BP Statistical Review of World Energy* (London: 2011); consumption of metals from Tim Jackson, *Prosperity Without Growth* (London: Earthscan, 2009), Chapter 5; logic of limiting throughput from Herman E. Daly, "Foreword," in ibid., pp. xi—ii; Chris Goodall, "'Peak Stuff' Did the UK Reach a Maximum Use of Material Resources in the Early Part of the Last Decade?" Research Paper, *Carboncommentary.com*, 13 October 2011; change in economic and social structures from Department of Economic and Social Affairs (DESA), *World Economic and Social Survey 2011, The Great Green Technological Transformation* (New York: United Nations, 2011); consumption by the wealthiest from World Bank, *World Bank Development Indicators 2008* (Washington, DC: 2008); Ricardo Abramovay, *A Transição para uma*

Nova Economia (São Paulo, Avina Foundation, November 2011).

23. Mark Halle, "Accountability in the Green Economy," in Selin and Najam, op. cit. note 1, p. 19.

24. Quote from G77 at Delhi Ministerial Dialogue on "Green Economy and Inclusive Growth," Inaugural Session, 3 October 2011, at www.uncsd 2012.org/rio20/content/documents/G77+China %20Discurso%20Del%20Embajador%20Al%20V arez%20En%20El.pdf. These concerns are also addressed in detail in *The Transition to a Green Economy: Benefits, Challenges and Risks from a Sustainable Development Perspective*, Report by a Panel of Experts to Second Preparatory Committee Meeting for United Nations Conference on Sustainable Development.

25. Figure 1–2 based on Global Footprint Network, *The Ecological Wealth of Nations* (Oakland, CA: 2010), pp. 28–35.

26. U.N. Development Programme, *Human Development Report 2011* (New York: Palgrave Macmillan, 2011), p. 68.

27. Herman E. Daly, ed., *Toward a Steady-State Economy* (San Francisco: W. H. Freeman & Co., 1973); on degrowth, see www.degrowth.org/What-is-Degrowth.22.0.html.

28. Saleemul Huq, "Climate and Energy," in Selin and Najam, op. cit. note 1; Alex Evans and David Steven, *Making Rio 2012 Work: Setting the Stage for Global Economic, Social and Ecological Renewal* (New York: Center on International Cooperation, New York University, 2011), p. 8.

29. Bigg, op. cit. note 1, p. 29.

30. Lack of access to sanitation from UNEP, op. cit. note 6, p. 19; urban water shortages from Foresight, *Migration and Global Environmental Change: Final Project Report* (London: The Government Office for Science, 2011), p. 191; International Energy Agency (IEA), *Energy For All: Financing Access for the Poor*, special early excerpt for the *World Energy Outlook 2011* (Paris: 2011); UN-Energy Knowledge Network, "Energy Access," at

www.un-energy.org/cluster/energy_access.

31. BP, op. cit. note 22.

32. UNEP, *Green Jobs*, op. cit. note 20.

33. Bloomberg New Energy Finance, "Clean Energy Investment Storms to New Record in 2010," press release (New York: 11 January 2011); REN21, *Renewables 2005 Global Status Report* (Washington, DC: 2005); Figure 1–3 adapted from Reference Tables in REN21, *Renewables 2011 Global Status Report* (Paris: 2011), pp. 71–72, 74–75.

34. REN21, *Renewables 2011*, op. cit. note 33, pp. 18–19.

35. Li Junfeng, Shi Pengfei, and Gao Hu, *2010 China Wind Power Outlook* (Beijing and Brussels: Chinese Renewable Energy Industries Association, Global Wind Energy Council, and Greenpeace, 2010); REN21, *Renewables 2011*, op. cit. note 33, pp. 15, 39; Olga Strietska-Ilina et al., *Skills for Green Jobs: A Global View—Synthesis Report Based on 21 Country Studies* (Geneva: ILO, 2011); J. M. Roig Aldasoro, "Navarre: Renewable Energies," Regional Minister of Innovation, Enterprise and Employment, Government of Navarre, Pamplona, 21 April 2009.

36. Solar power from REN21, *Renewables 2011*, op. cit. note 33, p. 41; solar cookers and solar lanterns from Lighting Africa, *Kenya: Qualitative Off-Grid Lighting Market Assessment* (Washington, DC: International Finance Corporation, 2008); Bangladesh from Strietska-Ilina et al., op. cit. note 35, and from Infrastructure Development Company Limited, "Progress with SHS's Installation up to 31 August 2011," at www.idcol.org/prjshsm2004 .php.

37. REN21, *Renewables 2011*, op. cit. note 33; Brazil from General Secretariat of the Presidency of the Republic, *The National Commitment to Improve Labor Conditions in the Sugarcane Activity* (Brasilia: undated).

38. Figure of 4.3 million is author estimate based on an extensive literature survey undertaken for a forthcoming ILO study on green jobs. It consists

of these rough job estimates: wind power, 670,000; solar PV, > 600,000; solar hot water, 870,000; biofuels, 1.5 million, and biomass, 600,000. The author also generated the 2008 figure for UNEP, *Green Jobs*, op. cit. note 20, p. 127.

39. Figure of 10 million is derived from an estimate of 3 million oil and gas extraction jobs and 7 million coal mining jobs. See ILO, "Promoting Decent Work in a Green Economy," ILO Background Note to UNEP, op. cit. note 6, and World Coal Institute, *The Coal Resource. A Comprehensive Overview of Coal* (London: 2005).

40. Emissions from UNEP, op. cit. note 6; vehicles from Colin Couchman, IHS Automotive, London, e-mail to author, 31 May 2011.

41. Fuel efficiency from UNEP, *Green Jobs*, op. cit. note 20; "Hybrid Car Statistics," undated, at www.all-electric-vehicles.com/hybrid-car-statistics .html.

42. Brazil from World Bank, *Brazil Low Carbon Country Case Study* (Washington, DC: 2010), and from Associação Nacional dos Fabricantes de Veiculos Automotores, *Anuário da Indústria Automobilística Brasileira*, Edition 2011 (São Paulo: 2011); compressed natural gas from International Association for Natural Gas Vehicles, "Natural Gas Vehicle Statistics," at www.iangv.org/tools-resources/ statistics.html, updated April 2011, and from NGV America, "NGVs and Biomethane," at www.ngvc .org/about_ngv/ngv_biomethane.html.

43. Motor vehicle employment from International Organization of Motor Vehicle Manufacturers, "Employment," at oica.net/category/economic -contributions/auto-jobs; rail vehicle manufacturing from Michael Renner and Gary Gardner, *Global Competitiveness in the Rail and Transit Industry* (Washington, DC: Worldwatch Institute, September 2010), p. 15; urban transit employment from International Association of Public Transport, "Employment in Public Transport: 13 Million People Worldwide!" April 2011; railway employment from International Union of Railways (UIC), *Rail and Sustainable Development* (Paris: 2011).

44. Renner and Gardner, op. cit. note 43; high-speed rail from UIC, *Km of High Speed Lines in the*

STATE OF THE WORLD 2012

World, at www.uic.org/IMG/pdf/20110701_b1_resume_km_of_hs_lines_in_the_world.pdf; Bus Rapid transit from Naoko Matsumoto, *Analysis of Policy Processes to Introduce Bus Rapid Transit Systems in Asian Cities from the Perspective of Lesson-drawing: Cases of Jakarta, Seoul, and Beijing* (Tokyo: Institute for Global Environmental Strategies, undated).

45. Building share of energy use from IEA, *World Energy Outlook 2010* (Paris: 2010); share of electricity use from IEA, *World Energy Outlook 2009* (Paris: 2009); projection from IEA and OECD, *Energy Technology Perspectives 2010 Scenarios and Strategies to 2050* (Paris: 2010).

46. UNEP, *Green Jobs*, op. cit. note 20, p. 131.

47. Slum inhabitants from UN-HABITAT, *State of the World's Cities 2010/2011: Bridging the Urban Divide* (London: Earthscan, 2010).

48. See, for instance, ILO, *Study of Occupational and Skill Needs in Green Building* (Geneva: 2011); Bracken Hendricks et al., *Rebuilding America: A National Policy Framework for Investment in Energy Efficiency Retrofits* (Washington, DC: Center for American Progress and Energy Future Coalition, 2009); Agence de l'Environnement et la Maîtrise de l'Energie (ADEME), *Activities Related to Renewable Energy and Energy Efficiency: Markets, Employment and Energy Stakes 2006–2007, Projections 2012* (Paris: 2008).

49. McGraw Hill Construction, *Green Building Retrofit and Renovation: Rapidly Expanding Market Opportunities through Existing Building* (New York: 2009).

50. See the website for the Directive, at www.epbd-ca.org; Meera Ghani, with Michael Renner and Ambika Chawla, *Low Carbon Jobs for Europe: Current Opportunities and Future Prospects* (Brussels: WWF, 2009), p. 23.

51. UNEP, *Global Green New Deal: An Update for the G20 Pittsburgh Summit* (Nairobi: 2009); Werner Schneider, "Green Jobs Creation in Germany," PowerPoint presentation at Cornell University, Institute for Labor Relations workshop on Climate Protection in Cities, States and Regions—Job Cre-

ation and Workforce Development, New York City, 12 May 2010; Political Economy Research Institute, *A New Retrofit Industry: An Analysis of the Job Creation Potential of Tax Incentives for Energy Efficiency in Commercial Buildings and Other Components of the Better Buildings Initiative* (Amherst, MA: University of Massachusetts, 2011).

52. Training programs from European Centre for the Development of Vocational Training, *Skills for Green Jobs: European Synthesis Report* (Luxembourg: Publications Office of the European Union, 2010); C. Martinez-Fernandez et al., *Greening Jobs and Skills Labour Market Implications of Addressing Climate Change* (Paris: OECD, 2010); Singapore Building and Construction Authority, *2nd Green Building Masterplan* (Singapore: 2009); India from ILO, op. cit. note 48, p. 111.

53. Extraction data from UNEP, *Recycling Rates of Metals: A Status Report* (Nairobi: 2011); waste collected from UNEP, op. cit. note 6.

54. Bureau of International Recycling, "Once Upon a Time … The Story of BIR, 1948–2008," at www.bir.org/assets/Documents/publications/brochures/BIRthday.pdf.

55. Institute for Local Self-Reliance, "Recycling Means Business," at www.ilsr.org/recycling/recyclingmeansbusiness.html; U.S. recycling employment from R. W. Beck, Inc., *U.S. Recycling Economic Information Study* (Washington, DC: National Recycling Coalition, 2001); EU employment from GHK, *Links Between the Environment, Economy and Jobs*, submitted to European Commission, DG Environment (London: 2007).

56. Collaborative Working Group (CWG) on Solid Waste Management in Low- and Middle-Income Countries and Deutsche Gesellschaft für Internationale Zusammenarbeit (GIZ), *The Economics of the Informal Sector in Solid Waste Management* (St. Gallen, Switzerland, and Eschborn, Germany: 2011).

57. Melanie Samson, "Introduction," in Melanie Samson, ed., *Refusing to be Cast Aside: Waste Pickers Organising Around the World* (Cambridge, MA: Women in Informal Employment: Globalizing and Organizing (WIEGO), 2009).

WWW.WORLDWATCH.ORG

58. The 1 percent and the 15 million estimate from Chris Bonner, "Waste Pickers Without Frontiers," *South African Labour Bulletin*, vol. 32, no. 4 (2008); additional calculation from Population Reference Bureau, *2011 World Population Data Sheet* (Washington, DC: 2011).

59. Martin Medina, "The Informal Recycling Sector in Developing Countries," *Gridlines*, October 2008; Sonia M. Dias, "Overview of the Legal Framework for Inclusion of Informal Recyclers in Solid Waste Management in Brazil," *WIEGO Urban Policies Briefing Note No. 8*, May 2011; Sonia M. Dias and F. C. G. Alves, *Integration of the Informal Recycling Sector in Solid Waste Management in Brazil* (Berlin: GTZ, 2008); "Brazilian President Launches the "Cata Ação" Program," *AVINA 2009 Annual Report*, at www.informeavina2009.org/english/reciclaje.shtml; "Brazil Sanctions National Policy that Formalizes the Work of 800,000 Recyclers," *AVINA 2010 Annual Report*, at www.informeavina2010.org/english/reciclaje.shtml.

60. CWG and GIZ, op. cit. note 56; WIEGO, "Waste Pickers," at wiego.org/informal-economy/occupational-groups/waste-pickers; WIEGO, "Laws & Policies Beneficial to Waste Pickers," at wiego.org/informal-economy/laws-policies-beneficial-waste-pickers.

61. Chris Bonner, "Foreword," in Samson, op. cit. note 57.

62. Box 1–2 based on the following: U.S.-China wind case from United Steelworkers' Section 301 Petition, from Jonathan Watts, "China Moves to Defuse Trade Row with US over Green Technology," (London) *Guardian*, 23 December 2010, from Doug Palmer and Leonora Walet, "China Agrees to Halt Subsidies to Wind Power Firms," *Reuters*, 7 June 2011, from Kevin Gallagher, "US Should Exercise Green Power," (London) *Guardian*, 6 January 2011, from Ed Crooks, "Washington's Energy Rift with China Unpopular," *Financial Times*, 19 October 2010, and from Dale Jiajun Wen, "Pointing the Finger the Wrong Way," *China Dialogue*, 19 October 2010; U.S.-China solar trade case from Keith Bradsher, "U.S. Solar Panel Makers Say China Violated Trade Rules," *New York Times*, 20 October 2011, from Keith Bradsher, "Chinese Trade Case Has Clear Targets,

Not Obvious Goals," *New York Times*, 21 October 2011, from James Kanter, "Trade Disputes Hurt Renewable Energy, Chinese Executive Asserts," *New York Times*, 27 October 2011, and from Stephen Lacey, "Chinese Cheaters? How China Dominates Solar," *Grist*, 10 September 2011; Japan-Ontario case from Todd Tucker, "Corporations Push for WTO Attack on Green Jobs," *Eyes on Trade*, Public Citizen's Global Trade Watch, 24 June 2011, from Gloria Gonzalez, "EU Joins Japan in Attacking Ontario's Renewables Tariff," *Environmental Finance*, 15 August 2011, from Shira Honig, "Japan Renewable Feed-in-Tariff Passes, While Ontario Faces Battles," *Climatico*, 7 September 2011, from John Landers, "Legal Issues for Ontario's Feed-in-Tariff Policies," *Energy Trend*, 18 August 2011, and from Paul Gipe, "Japan Feed-in Tariff Policy Becomes Law," *Wind-Works*, 27 August 2011.

63. DESA, op. cit. note 22, pp. xx–xxi, with detailed discussion in Section VI (pp. 161–86).

64. Energy Conservation Center, at www.eccj.or.jp/top_runner/index.html. The World Economic and Social Survey discusses Japan's experience and suggests a global top runner program; see DESA, op. cit. note 22, pp. 47, 61.

65. "Increasing Price with Volume," in UNEP and GRID-Arendal, *Vital Water Graphics*, 2nd ed. (Arendal, Norway: 2008).

66. David Schweickart, "A New Capitalism—or a New World?" *World Watch*, September/October 2010; Tom Prugh, "Band-aids for Capitalism? Or Something Completely Different?" (blog), *Worldwatch Green Economy*, 28 August 2009.

Chapter 2. The Path to Degrowth in Overdeveloped Countries

1. Number of participants from "Degrowth Conference Barcelona 2010," at degrowth.eu. Box 2–1 from the following: Serge Latouche, "Growing a Degrowth Movement," in Worldwatch Institute, *State of the World 2010* (New York: W. W. Norton & Company, 2010), p. 181; Serge Latouche, *Farewell to Growth* (Cambridge, U.K.: Polity Press, 2009), pp. 8–9; Tim Jackson, *Prosperity Without*

Growth: Economics for a Finite Planet (London: Earthscan, 2011).

2.   Martín Mucha, "Robin Bank, Héroe Juvenil," *El Mundo,* 12 October 2010; Erik Assadourian, "A Tale of a Modern-Day Robin Hood" (blog) *Transforming Cultures,* 4 November 2010; Enric Duran, "I Have 'Robbed' 492,000 Euros to Whom Most Rob Us in Order to Denounce Them and Build Some Alternatives for the Society" (blog) *enricduran.cat,* 27 August 2010; Giles Tremlett, "€500,000 Scam of a Spanish Robin Hood," (London) *Guardian,* 18 September 2008.

3.   Millennium Ecosystem Assessment, *Living Beyond Our Means: Natural Assets and Human Well-Being: Statement from the Board* (Washington, DC: World Resources Institute, 2005), p. 2; Johan Rockström et al., "A Safe Operating Space for Humanity," *Nature,* 24 September 2009, pp. 472–75.

4.   Andrei Sokolov et al., "Probabilistic Forecast for 21st Century Climate Based on Uncertainties in Emissions (without Policy) and Climate Parameters," *American Meteorological Society Journal of Climate,* October 2009, pp. 5,175–204; David Chandler, "Revised MIT Climate Model Sounds Alarm," *TechTalk* (Massachusetts Institute of Technology), 20 May 2009; Juliet Eilperin, "New Analysis Brings Dire Forecast of 6.3-Degree Temperature Increase," *Washington Post,* 25 September 2009; Elizabeth R. Sawin et al., "Current Emissions Reductions Proposals in the Lead-up to COP-15 Are Likely to Be Insufficient to Stabilize Atmospheric CO_2 Levels: Using C-ROADS—a Simple Computer Simulation of Climate Change—to Support Long-Term Climate Policy Development," draft presented at the Climate Change—Global Risks, Challenges, and Decisions Conference, University of Copenhagen, Denmark, 10 March 2009; Mark G. New et al., eds., "Four Degrees and Beyond: The Potential for a Global Temperature Increase of Four Degrees and Its Implications," *Philosophical Transactions of the Royal Society A,* 13 January 2011; "Royal Society Special Issue Details 'Hellish Vision' of 7°F (4°C) World—Which We May Face in the 2060s!" *Climate Progress,* 29 November 2010; Richard Black, "Climate Talks End with Late Deal," *BBC News,* 11 December 2011.

5.   "Canada to Withdraw from Kyoto Protocol," *BBC News,* 13 December 2011; Fiona Harvey, "Rich Nations 'Give Up' on New Climate Treaty Until 2020," (London) *Guardian,* 20 November 2011; Executive Summary, *The Economics of Climate Change: The Stern Review* (Cambridge, U.K.: Cambridge University Press, 2007), p. 10.

6.   Harald Welzer, *Mental Infrastructures: How Growth Entered the World and Our Souls* (Berlin: Heinrich Böll Foundation, 2011), p. 12; "Prince of Wales: Ignoring Climate Change Could Be Catastrophic," (London) *Telegraph,* 24 May 2011.

7.   Welzer, op. cit. note 6, p. 10; WWF, ZSL, and GFN, *Living Planet Report 2010* (Gland, Switzerland: 2010); World Bank, "New Data Show 1.4 Billion Live On Less Than US$1.25 A Day, But Progress Against Poverty Remains Strong," press release (Washington, DC: 26 August 2008).

8.   Overweight Americans from Trust for America's Health, *F as in Fat: How Obesity Policies Are Failing in America* (Washington, DC: Robert Wood Johnson Foundation, 2008); medical and productivity costs from Society of Actuaries, "New Society of Actuaries Study Estimates $300 Billion Economic Cost Due to Overweight and Obesity," press release (Schaumburg, IL: 10 January 2011), and from Robert Preidt, "Cost of Obesity Approaching $300 Billion a Year," *USA Today,* 12 January 2011; Institute for Health Metrics and Evaluation, "Life Expectancy in Most US Counties Falls Behind World's Healthiest Nations," press release (Seattle, WA: 15 June 2011); David Brown, "Life Expectancy in the U.S. Varies Widely by Region, in Some Places Is Decreasing," *Washington Post,* 15 June 2011; S. Jay Olshansky et al., "A Potential Decline in Life Expectancy in the United States in the 21st Century," *New England Journal of Medicine,* 17 March 2005, pp. 1,138–45; Laura Cummings, "The Diet Business: Banking on Failure," *BBC News,* 5 February 2003; global obesity from Richard Weil, "Levels of Overweight on the Rise," *Vital Signs Online,* 14 June 2011.

9.   Juliet Schor, "Sustainable Work Schedules for All," in Worldwatch Institute, op. cit. note 1, pp. 91–95; Gary Gardner, Erik Assadourian, and Radhika Sarin, "The State of Consumption Today," in Worldwatch Institute, *State of the World 2004* (New

York: W. W. Norton & Company, 2004), pp. 3–21; Sonia Shah, "As Pharmaceutical Use Soars, Drugs Taint Water and Wildlife," *Yale Environment 360*, 15 April 2010; Miller McPherson, Lynn Smith-Lovin, and Matthew E. Brashears, "Social Isolation in America: Changes in Core Discussion Networks over Two Decades," *American Sociological Review*, June 2006, pp. 353–75.

10. Sustainable Europe Research Institute, GLOBAL 2000, and Friends of the Earth Europe, *Overconsumption? Our Use of the World's Natural Resources* (September 2009).

11. Zenith Optimedia, "Quadrennial Events to Help Ad Market Grow in 2012 Despite Economic Troubles," press release (London: 5 December 2011); Jack Neff, "Is Digital Revolution Driving Decline in U.S. Car Culture?" *Advertising Age*, 31 May 2010; Lisa Hymas, "Driving Has Lost Its Cool for Young Americans," *Grist*, 27 December 2011; Victoria J. Rideout, Ulla G. Foehr, and Donald F. Roberts, *Generation M2: Media in the Lives of 8- to 18-Year-Olds* (Washington, DC: Kaiser Family Foundation, 2010).

12. Erik Assadourian, "The Rise and Fall of Consumer Cultures," in Worldwatch Institute, op. cit. note 1, pp. 3–20; Paul Taylor and Wendy Wang, "The Fading Glory of the Television and Telephone," Pew Research Center, Washington, DC, 19 August 2010. Box 2–2 is based on Michael Maniates and John M. Meyer, eds., *The Environmental Politics of Sacrifice* (Cambridge, MA: The MIT Press, 2010).

13. Michael Maniates, "Editing Out Unsustainable Behavior," in Worldwatch Institute, op. cit. note 1, pp. 119–26; Brian Merchant, "Plastic Bags Used in DC Drop From 22 Million to 3 Million a Month," *Treehugger*, 31 March 2010; "Good News, Bad News on D.C.'s Plastic Bag Tax," *Washington Examiner*, 5 January 2011.

14. Sheryl Gay Stolberg, "Wal-Mart Shifts Strategy to Promote Healthy Foods," *New York Times*, 20 January 2011; Bruce Blythe, "UPDATED: Wal-Mart's Health Kick Cuts Prices on Produce," *The Packer*, 20 January 2011; Tom Philpott, "Is Wal-Mart Our Best Hope for Food Policy Reform?" *Grist*, 29 April 2011.

15. Adam Aston, "Patagonia Takes Fashion Week as a Time to Say: 'Buy Less, Buy Used,'" *GreenBiz*, 8 September 2011; Tim Nudd, "Ad of the Day: Patagonia," *Ad Week*, 28 November 2011.

16. Nudd, op. cit. note 15.

17. David Reay, *Climate Change Begins at Home* (New York: MacMillan, 2005); estimate from National Funeral Directors Association, "Statistics: Funeral Costs," at www.nfda.org/media-center/statisticsreports.html, viewed 28 December 2011, and from Selena Maranjian, "How Much Does a Funeral Cost?" *Fool.com*, 5 March 2002; Joe Schee, "Presentation: Eco-Friendly End of Life Rituals," Green Burial Council, 2010.

18. Helene Gallis, "The Slow Food Movement," from Worldwatch Institute, op. cit. note 1, p. 182.

19. The Meatless Monday Campaign, "The Movement Goes Global," at www.meatlessmonday.com/the-movement-goes-global; Marc Gunther, "Sodexo's Meatless Mondays Give 'Where's the Beef' a New Meaning," *GreenBiz*, 2 May 2011.

20. "What's Cooking, Uncle Sam?" exhibit at U.S. National Archives, viewed 16 December 2011; Erik Assadourian, "Uncle Sam Says Garden...And Eat Vitamin Donuts" (blog), *Transforming Cultures*, 17 December 2011.

21. Calculation based on Global Footprint Network, *The Ecological Footprint Atlas 2008* (Oakland, CA: rev. ed., 16 December 2008); G. Ananthapadmanabhan, K. Srinivas, and Vinuta Gopal, *Hiding Behind the Poor* (Bangalore: Greenpeace India Society, 2007); Assadourian, op. cit. note 12.

22. Richard Wilkinson and Kate Pickett, *The Spirit Level: Why More Equal Societies Almost Always Do Better* (London: Penguin Group, 2009).

23. U.N. Development Programme, "2011 Human Development Index Covers Record 187 Countries and Territories, Puts Norway at Top, DR Congo Last," press release (Copenhagen: 2 November 2011).

24. "U.S. Federal Individual Income Tax Rates

History, 1913–2011 (Nominal and Inflation-Adjusted Brackets)," Tax Foundation, Washington, DC, 9 September 2011.

25. Steven Greenhouse and Graham Bowley, "Tiny Tax on Financial Trades Gains Advocates," *New York Times*, 6 December 2011.

26. James Grubel, "Australia Passes Landmark Carbon Price Laws," *Reuters*, 8 November 2011; Enda Curran and Ray Brindal, "Australia's Carbon Tax Clears Final Hurdle," *Wall Street Journal*, 8 November 2011; "Australia Makes Green Cuts to Fund Flood Relief," *Radio Australia*, 28 January 2011.

27. Zenith Optimedia, op. cit. note 11; Zoe Gannon and Neal Lawson, *The Advertising Effect: How Do We Get the Balance of Advertising Right* (London: Compass, 2010).

28. Mark Hertsgaard, *Hot: Living Through the Next Fifty Years on Earth* (New York: Houghton Mifflin Harcourt, 2011), pp. 107–27.

29. Ibid.; Jason Samenow, "NOAA: 2011 Sets Record for Billion Dollar Weather Disasters in the U.S.," *Washington Post*, 7 December 2011; Petra Löw, "Losses From Natural Disasters Decline in 2009," *Vital Signs Online*, 25 March 2010.

30. Anna Coote, Jane Franklin, and Andrew Simms, *21 Hours: Why a Shorter Working Week Can Help Us All to Flourish in the 21st Century* (London: New Economics Foundation, 2010).

31. Tim Kasser and Kirk Brown, as cited in Juliet Schor, *Plenitude: The New Economics of True Wealth* (New York: Penguin Press, 2010), pp. 113–14, and 178; Gary Gardner and Erik Assadourian, "Rethinking the Good Life," in Worldwatch Institute, op. cit. note 9.

32. Juliet Schor, *The Overworked American: The Unexpected Decline of Leisure* (New York: Basic Books, 1993); Netherlands from John de Graff, "Reducing Work Time as a Path to Sustainability," in Worldwatch Institute, op. cit. note 1, pp. 173–77; "Employers and Unions Brace for a Downturn," *Der Spiegel*, 20 October 2011; Christian Vits and

Jana Randow, "The Price of Saving Jobs in Germany," *Business Week*, 29 July 2010; Nicholas Kulish, "Aided by Safety Nets, Europe Resists Stimulus Push," *New York Times*, 26 March 2009.

33. De Graff, op. cit. note 32; Michael Maniates, "Struggling with Sacrifice: Take Back Your Time and Right2Vacation.org," in Maniates and Meyer, op. cit. note 12, pp. 293–312; Sweden from Organisation for Economic Co-operation and Development, "PF2.1: Key Characteristics of Parental Leave Systems," 15 April 2011, at www.oecd.org/dataoecd/45/26/37864482.pdf, pp. 6 and 14. Note: Sweden has a ceiling of 43,070 euros for parental leave.

34. Juliet Schor, *The Overspent American: Why We Want What We Don't Need* (New York: Harper Perennial, 1999); Schor, op. cit. note 32; Schor, op. cit. note 31.

35. Rakesh Kochhar and D'Vera Cohn, *Fighting Poverty in a Bad Economy, Americans Move in with Relatives* (Washington, DC: Pew Research Center, 2011); Catherine Rampell, "As New Graduates Return to Nest, Economy Also Feels the Pain," *New York Times*, 16 November 2011.

36. Jessica Silver-Greenberg, "When Kids Come Back Home, *Wall Street Journal*, 26 November 2011.

37. Beth Snyder Bulik, "Boom in Multigenerational Households Has Wide Implications for Ad Industry," *Advertising Age*, 23 August 2010.

38. "What's Cooking, Uncle Sam?" op cit. note 20; Peter Rosset and Medea Benjamin, *Two Steps Backward, One Step Forward: Cuba's Nationwide Experiment with Organic Agriculture* (San Francisco: Global Exchange, 1993); The Community Solution, *The Power of Community: How Cuba Survived Peak Oil* (Yellow Springs, OH: 2006); Mario Gonzalez Novo and Catherine Murphy, "Urban Agriculture in the City of Havana: A Popular Response to a Crisis," in N. Bakker et al., eds., *Growing Cities Growing Food: Urban Agriculture on the Policy Agenda: A Reader on Urban Agriculture* (German Foundation for International Development, 2001), pp. 329–47.

39. Juliet Schor, "Exit Ramp to Sustainability: Building a Small-scale, Low-footprint, High-knowledge Economy," presentation at SCORAI workshop, Princeton, NJ, 16 April 2011; Cecile Andrews and Wanda Urbanska, "Inspiring People to See Less Is More," in Worldwatch Institute, op. cit. note 1, pp. 178–84; Shareable.net: Sharing By Design, at shareable.net; Gardner and Assadourian, op. cit. note 31.

40. Nicole Winfield, "Pope Laments Christmas Consumerism, Urges People to Look Beyond 'Superficial Glitter,'" *Huffington Post*, 24 December 2011; Gary Gardner, "Engaging Religions to Shape Worldviews," in Worldwatch Institute, op. cit. note 1, pp. 23–29; Gary Gardner, "Ritual and Taboo as Ecological Guardians," in Worldwatch Institute, op. cit. note 1, pp. 30–35; St. Francis Pledge from Catholic Climate Covenant, at catholicclimatecovenant.org.

41. Chuck Collins, presentation on Common Security Clubs, Washington, DC, 18 January 2011; Common Security Clubs website, at localcircles.org.

42. Transition Network website, at www.transitionnetwork.org/initiatives/map; Andrews and Urbanska, op. cit. note 39; Shaftesbury Transition Town website, at www.transitiontownshaftesbury.org.uk.

43. David Orr, "The Oberlin Project: What Do We Stand for Now?" *Oberlin Alumni Magazine*, fall 2011.

44. Erik Assadourian, "Sustainable Communities Become More Popular," *Vital Signs 2007–2008* (New York: W. W. Norton & Company, 2007), pp. 104–05; Jonathan Dawson, "Ecovillages and the Transformation of Values," in Worldwatch Institute, op. cit. note 1, pp. 185–90.

45. Jennifer Block, *Pushed: The Painful Truth About Childbirth and Modern Maternity Care* (Philadelphia: De Capo Press, 2007); Steven Reinberg, "C-section Rate in U.S. Climbs to All-Time High," *USA Today*, 22 July 2011; Jennifer Block, "Midwife Q&A: Are We Having Babies All Wrong?" *Time*, 25 May 2011; A. Mark Durand,

"The Safety of Home Birth: The Farm Study," *American Journal of Public Health*, March 1992, pp. 450–52.

46. Kevin Green and Erik Assadourian, "Making Social Welfare Programs Sustainable," in Worldwatch Institute, op. cit. note 1, p. 141; Francesco di Iacovo, "Social Farming: Dealing with Communities Rebuilding Local Economy," presentation at Rural Futures Conference, University of Plymouth, U.K., 1–4 April 2008.

47. Rachel Donadio, "With Work Scarce in Athens, Greeks Go Back to the Land," *New York Times*, 8 January 2012.

48. Richard E. White and Gloria Eugenia González Mariño, "Las Gaviotas: Sustainability in the Tropics," *World Watch Magazine*, May/June 2007, pp. 18–23; Friends of Gaviotas website, at www.friendsofgaviotas.org.

49. Jackson, op. cit. note 1, p. 185.

50. Assadourian, op. cit. note 12, p. 12.

51. New Economics Foundation, *The Impossible Hamster*, Script: Andrew Simms, Animators: Leo Murray and Thomas Bristow, London, 2010.

52. Latouche, "Growing a Degrowth Movement," op. cit. note 1; see also Degrowthpedia, at degrowthpedia.org.

53. Kick It Over! website, at www.kickitover.org; Kick It Over Manifesto, at kickitover.org/sites/default/files/downloads/adb_poster_manifesto.pdf; Michael C. George, "Group Endorses Walk Out in Economics 10," *Harvard Crimson*, 2 November 2011; "An Open Letter to Greg Mankiw," *Harvard Political Review*, 2 November 2011; Net Impact from Erik Assadourian, "Maximizing the Value of Professional Schools," in Worldwatch Institute, op. cit. note 1, p. 78; Net Impact website, at netimpact.org.

54. Worldwatch Institute, "Oil Discovered on the Island of Catan," press release (Washington, DC: 19 October 2011); rules of *Catan Scenarios: Oil Springs* at www.oilsprings.catan.com.

Chapter 3. Planning for Inclusive and Sustainable Urban Development

1. B. Sanyal, "Planning as Anticipation of Resistance," *Planning Theory*, vol. 4, no. 3 (2005), pp. 225–45; J. L. Baker and K. McClain, *Private Sector Initiatives in Slum Upgrading*, Urban Papers (Washington, DC: World Bank, 2009); for an asset-building framework, see C. O. N. Moser, *Asset-based Approaches to Poverty Reduction in a Globalized Context*, Brookings Global Economy and Development Working Paper (Washington, DC: Brookings Institution, 2006).

2. Table 3–1 from Population Division, *World Urbanization Prospects: The 2009 Revision* (New York: United Nations, 2009).

3. Box 3–1 from the following: megacity sizes from U.N. Environment Programme, *Keeping Track of Our Changing Environment: From Rio to Rio+20 (1992–2012)* (Nairobi: 2011), from UN-HABITAT, *State of the World's Cities 2008/2009: Harmonious Cities* (London: United Nations, 2008), and from Population Division, *World Urbanization Prospects: The 2007 Revision* (New York: United Nations, 2007); energy use from UN-HABITAT, Cities and Climate Change Initiative Launch and Conference Report, Oslo, 2009; energy opportunities from World Bank, *State and Trends of the Carbon Market 2010* (Washington, DC: 2010); drinking water from UN-Water Decade Programme on Advocacy and Communication, "Water and Cities: Facts and Figures," at www.un.org/waterforlifedecade/; Delhi example from J. Pittock et al., *Interbasin Water Transfers and Water Scarcity in a Changing World—A Solution or a Pipedream?* (Frankfurt: WWF Germany, 2009); Dhaka from M. Sinha, "Community-based Waste Management and Composting for Climate/Co-benefits—Case of Bangladesh," presented at the International Consultative Meeting on Expanding Waste Management Services in Developing Countries, Tokyo, 18–19 March 2010; opportunities for mitigation and adaptation from United Nations, "The Challenge of Adapting to a Warmer Planet for Urban Growth and Development," UN-DESA Policy Brief No. 25, New York, December 2009, and from UN-HABITAT, *State of the World's Cities 2008/2009*, op. cit. this note; United Nations, *Shanghai Manual: A Guide for Sustainable Urban Development in the 21st Century* (New York: 2011). For link between urbanization and income growth, see David E. Bloom and Tarun Khanna, "The Urban Revolution," *Finance and Development*, September 2007; additional data from UN-HABITAT, *State of the World's Cities 2010/2011: Cities for All* (London: Earthscan, 2010).

4. UN-HABITAT, *State of the World's Cities 2010/2011*, op. cit. note 3.

5. Ibid.

6. Michael Majale, "Employment Creation through Participatory Urban Planning and Slum Upgrading," *Habitat International*, vol. 32, no. 2 (2008), pp. 270–82; International Labour Organization, *Statistical Update on Employment in the Informal Economy* (Geneva: June 2011); C. K. Pralahad and Allen Hammond, "Serving the World's Poor, Profitably," *Harvard Business Review*, September 2002.

7. Pralahad and Hammond, op. cit. note 6; McKinsey Global Institute, *India's Urban Awakening: Building Inclusive Cities, Sustaining Economic Growth* (McKinsey & Company, 2010).

8. UN-HABITAT, *Forced Evictions—Towards Solutions?* (Nairobi: U.N.-Habitat Advisory Group on Forced Evictions, 2007); J. L. Baker and K. McClain, *Private-Sector Initiatives in Slum Upgrading* (Washington, DC: World Bank, 2009).

9. UN-HABITAT, *Global Report on Human Settlements 2009: Planning Sustainable Cities* (London: United Nations, 2009), pp. 153–54.

10. Ibid.

11. McKinsey Global Institute, op. cit. note 7; UN-HABITAT, *Slums of the World: The Face of Urban Poverty in the New Millennium?* (Nairobi: 2003).

12. UN-HABITAT, op. cit. note 9, p. 26; Vanessa Watson, "'The Planned City Sweeps the Poor Away...': Urban Planning and 21st Century Urbanisation," *Progress in Planning*, vol. 72, no. 3 (2009), p. 177; UN-HABITAT, op. cit. note 11.

13. UN-HABITAT, *Housing the Poor in Asian Cities: Community-Based Organizations—The Poor as Agents of Development* (Nairobi: 2008).

14. Sanyal, op. cit. note 1.

15. World Bank, *World Development Report 1997: The State in a Changing World* (Washington, DC: 1997).

16. Ibid.

17. Sanyal, op. cit. note 1; UN-HABITAT, *Secure Land Rights for All* (Nairobi: 2008).

18. Ministry of Urban Employment and Poverty Alleviation and Ministry of Urban Development, *Jawaharlal Nehru National Urban Renewal Mission: Overview* (Delhi: Government of India, undated); other examples from World Bank, *World Development Report 2009: Reshaping Economic Geography* (Washington, DC: 2009), Chapter 7.

19. Gerald Frug, "Governing the Megacity," Urban Age, Mexico City, 2006; UN-HABITAT, op. cit. note 9; Pearl River Delta from World Bank, op. cit. note 18.

20. Francos Halla, "Preparation and Implementation of a General Planning Scheme in Tanzania: Kahama Strategic Urban Development Planning Framework," *Habitat International*, vol. 26, no. 2 (2002), pp. 281–93.

21. United Nations, *Shanghai Manual*, op. cit. note 3.

22. World Bank, *Eco2 Cities: Ecological Cities as Economic Cities* (Washington, DC: 2009).

23. Aprodicio Laquin, discussion with author, Cambridge, MA, 27 May 2011, citing Liangyong Wu, *Rehabilitating the Inner City of Beijing, A Project in the Ju'erHutong Neighborhood* (Vancouver: University of British Columbia Press, 1999), Jeff Kenworthy, "Urban Ecology in Indonesia: The Kampung Improvement Program," Murdoch University, 1997, at www.istp.murdoch.edu.au/ISTP/casestudies/Case_Studies_Asia/kip/kip.html, and Ayse Pamuk and P. Cavallieri, "Alleviating Urban Poverty in a Global City: New Trends in Upgrading Rio de Janeiro's Favelas," *Habitat International*, vol. 22, no. 4 (1998), pp. 449–62.

24. Mary Schmidt, "Popular Participation and the World Bank: Lessons from Forty-Eight Case Studies," in J. Rietbergen-McCracken, ed., *Participation in Practice: The Experience of the World Bank and Other Stakeholders*, Discussion Paper No. 333 (Washington, DC: World Bank, 1996), pp. 21–25; Kamla Raheja Vidyanidhi Institute for Architecture and Environmental Studies and the Society for the Promotion of Area Resource Centres, *Interpreting, Imagining, Developing Dharavi* (Mumbai: 2010).

25. E. Hagen, *Putting Nairobi's Slums on the Map* (Washington, DC: World Bank Institute, 2010); Global Land Tool Network Secretariat, "Improving Data Collection for Urban Planning through Participatory Enumerations," Global Land Tool Network Brief, UN-HABITAT, Nairobi, 2010.

26. Porto Alegre from United Nations, *Shanghai Manual*, op. cit. note 3, pp. 29–33.

27. UN-HABITAT, op. cit. note 9.

28. S. Patel et al., "Beyond Evictions in a Global City: People-managed Resettlement in Mumbai," *Environment and Urbanization*, vol. 14, no. 1 (2002), pp. 159–72; World Bank, op. cit. note 22; UN HABITAT, *Land and National Disasters: Guidance for Practitioners* (Nairobi: 2010).

29. Ketelan from *Slum Upgrading Facility Newsletter* (UN HABITAT), April 2008; Durban from N. R. Peirce and C. W. Johnson, *Century of the City: No Time to Lose* (New York: Rockefeller Foundation, 2009), pp. 132–39.

30. Alfonso X. Iracheta, "Evaluacion del Fondo Metropolitano, 2006–2009," Secretaria de Hacienda y Credito Publico and Inter-American Development Bank, El Colegio Mexiquense, 2010, at www.transparenciapresupuestaria.gob.mx.

31. For a brief overview, see D. Satterthwaite, "Getting Land for Housing; What Strategies Work for Low-Income Groups?" *Environment and Urbanization*, October 2009, pp. 299–307, and E. Fernandes, *Regularization of Informal Settlements* *in Latin America: Policy Focus Report* (Cambridge, MA: Lincoln Institute of Land Policy, 2011).

32. For an overview of these issues in practice, see UN-HABITAT, op. cit. note 11, and UN-HABITAT, op. cit. note 9.

33. For an in-depth discussion of these issues, see World Bank, op. cit. note 18, and World Bank, op. cit. note 22.

34. Cities Alliance, at www.citiesalliance.org/ca/sites/citiesalliance.org.

35. Manila from Asian Development Bank, *Bringing Water to the Poor, Selected ADB Case Studies* (Kyoto: 2004); A. Segel, M. Chu, and G. Herrero, "Patrimonio Hoy, Harvard Business School Case Study," Harvard Business School, Cambridge, MA, 2004.

Chapter 4. Moving Toward Sustainable Transport

1. "A Child Is Born and World Population Hits 7 Billion," MSNBC.com, 31 October 2011; World Health Organization (WHO), *Global Status Report on Road Safety* (Geneva: 2009); Health Effects Institute, *Traffic-Related Air Pollution: A Critical Review of the Literature on Emissions, Exposure, and Health Effects*, Special Report 17 (Boston: 2010).

2. Projections of auto numbers and reduction of greenhouse gases from International Energy Agency (IEA), *Transport, Energy and CO₂: Moving Towards Sustainability* (Paris: 2009); road fatality projections from WHO, op. cit. note 1; premature deaths from WHO, "Air Quality and Health," Fact Sheet No. 313, Geneva, September 2011, and from Health Effects Institute, op. cit note 1.

3. Enrique Peñalosa, "Urban Transport and Urban Development: A Different Model," presented at Center for Latin American Studies, University of California–Berkeley, 8 April 2002.

4. Emissions of global greenhouse gases from transport from IEA, op. cit. note 2; Institute for Transportation and Development Policy (ITDP), unpublished research on climate mitigation funds.

5. Walter Hook, "Urban Transport and the Millennium Development Goals," *Global Urban Development Magazine*, March 2006.

6. Ramon Cruz, ITDP, discussion with authors, New York, 29 December 2011.

7. Bridging the Gap, at www.bridgingthegap.org; Partnership for Sustainable Low-Carbon Transportation, at www.slocat.net.

8. Clean Air Portal, "Bangkok 2020 Declaration: Sustainable Transport Goals 2010–2020," August 2010, at cleanairinitiative.org/portal/node/6445; Foro de Transporte Sostenible para América Latina, "Bogota Declaration: Sustainable Transport Objectives," June 2011, at www.uncrdlac.org/fts/BogotaDeclaration.pdf; Report of the Secretary-General, *Policy Options and Actions for Expediting Progress in Implementation: Transport* (New York: United Nations, 2011).

9. IEA, op. cit. note 2; Joyce Dargay, Dermot Gately, and Martin Sommer, "Vehicle Ownership and Income Growth, Worldwide: 1960–2030," *Energy Journal*, vol. 28, no. 4 (2007), pp. 143–70; Figure 4–1 from IEA, op. cit. note 2.

10. U.N. Environment Programme, *Towards a Green Economy: Pathways to Sustainable Development and Poverty Eradication* (Nairobi: 2011), p. 378.

11. Air pollutants from ibid.

12. Ernesto Sanches-Triana et al., *Environmental Priorities and Poverty Reduction: A Country Environmental Analysis for Colombia* (Washington, DC: World Bank, 2007); U.S. Federal Highway Administration, *Addendum to the 1997 Federal Highway Cost Allocation Study* (Washington, DC: 2000); U.S. Environmental Protection Agency, *The Benefits and Costs of the Clean Air Act, 1970 to 1990* (Washington, DC: 1997).

13. WHO, *Night Noise Guideline for Europe* (Geneva: 2009).

14. D. Shrank and T. Lomax, *2011 Urban Mobility Report* (College Station, TX: Texas Transportation Institute, 2011); Standing Advisory Committee on Trunk Road Assessment, *Trunk Roads and the*

Generation of Traffic (London: U.K. Department of Transport, 1994); The Telegraph Business Club and IBM, *Future Focus: Travel* (London: 2009); U.N. Economic and Social Commission for Asia and the Pacific, U.N. Economic Commission for Latin America and the Caribbean, and Urban Design Lab, *Are We Building Competitive and Liveable Cities?* (2010).

15. "A Child Is Born," op. cit. note 1; IEA, op. cit. note 2.

16. Social Exclusion Unit, *Making the Connections: Final Report on Transport and Social Exclusion* (London: U.K. Government, 2003); P. Rode et al., *Cities and Social Equity: Inequality, Territory and Urban Form* (London: Urban Age Programme, London School of Economics, 2009); World Bank, *Cities on the Move: A World Bank Transport Strategy Review* (Washington, DC: 2002).

17. WHO, *World Traffic Safety Report* (Geneva: 2009).

18. Data and Figure 4–2 from ibid.

19. Ibid.; G. Jacobs and A. Aeron Thomas, "A Review of Global Road Accident Fatalities," presented at RoSPA Road Safety Congress, Plymouth, U.K., 3–7 March 2000; Hook, op. cit. note 5; Leonard J. Paulozzi et al., "Economic Development's Effect on Road Transport-related Mortality Among Different Road Users: A Cross-sectional International Study," *Accident Analysis & Prevention*, May 2007, pp. 606–17.

20. Intergovernmental Panel on Climate Change, *Climate Change 2007: Fourth Assessment Report* (Cambridge, U.K.: Cambridge University Press, 2007); Ben Block, "Interview: James Hansen Talks about Climate Change," *World Watch Magazine*, July/August 2008; IEA, *Energy Technology Perspectives 2008* (Paris: 2008).

21. Data and Figure 4–3 from IEA, op. cit. note 20.

22. Ibid., p. 425; Urban Land Institute (ULI), *Growing Cooler: The Evidence on Urban Development and Climate Change* (Washington, DC: 2007); ULI, *Moving Cooler* (Washington, DC: 2009).

23. Box 4–2 from ITDP, *Our Cities Ourselves* (New York: 2011).

24. K. Sakomoto, H. Dalkmann, and D. Palmer, *A Paradigm Shift Towards Sustainable Low-Carbon Transport* (New York: ITDP, 2010).

25. IEA, *Global Fossil Fuel Subsidies and the Impacts of Their Removal* (Paris: 2011); Global Subsidy Institute, *Joint Submission to the UN Conference on Sustainable Development, Rio+20* (Geneva: 2011).

26. ITDP, op. cit. note 4.

27. Table 4–2 from Stefan Bakker and Cornie Huizenga, "Making Climate Instruments Work for Sustainable Transport in Developing Countries," *Natural Resources Forum*, November 2010, pp. 314–26.

28. ITDP, unpublished research in budgets and project databases of all multilateral development banks, 2006–10.

29. Ibid.

30. Asian Development Bank, *Sustainable Transport Initiative Operational Plan* (Manila: 2010)

31. This proposal summarizes key elements from an official submission to the Rio+20 Conference from the Partnership for Sustainable Low Carbon Transport (over 50 nongovernmental organizations, multilateral development banks, associations, and agencies). For the complete submission, including target indicators, see "Partnership on Sustainable, Low Carbon Transport," Rio+20 website, at www.uncsd2012.org/rio20/index.php?page=view &type=510&nr=241&menu=20.

Chapter 5. Information and Communications Technologies Creating Livable, Equitable, Sustainable Cities

1. Singapore-MIT Alliance for Research and Technology, at smart.mit.edu/research/future-urban -mobility/research-projects.html; IBM, "IBM and Singapore's Land Transport Authority Pilot Innovative Traffic Prediction Tool," press release (Armonk,

NY: 1 August 2007); Lagos from West African NGO Network, "Habitat 2011 beta Physical Asset Tracking System (P.A.T.S)," at 1.latest.habitat-2011.appspot.com; India from Next Drop, at www.nextdrop.org; Figure 5–1 from International Telecommunication Union (ITU), "World Telecommunication/ICT Indicators Database," at www.itu.int/ITU-D/ict/publications/world/world.html.

2. Urbanization figure from World Bank, "Systems of Cities: Harnessing Urbanization for Growth and Poverty Alleviation," at World Bank Urban Strategy website; penetration in developing countries and Figure 5–2 from ITU, op. cit. note 1; Telecom Regulatory Authority of India, "Highlights of Telecom Subscription Data as on 31st October, 2011," press release (New Delhi: 8 December 2011).

3. Quote from Michel St. Pierre, "Sustainable Cities—Shanghai Looking Forward," presentation at CTBUH 2010, Mumbai, 3–5 February 2010.

4. Steven Erlanger, "A New Fashion Catches On in Paris: Cheap Bicycle Rentals," *New York Times*, 13 July 2008.

5. "Smart+Connected Communities," at www.cisco.com/web/strategy/smart_connected_communities.html; "Building Sustainable Cities," at www.ge-cities.com/en_GB; "Sustainable Cities," at www.usa.siemens.com/sustainable-cities.

6. Greg Lindsay, "Not-So Smart Cities," *New York Times*, 25 September 2011; Carlo Ratti and Anthony Townsend, "The Social Nexus," *Scientific American*, September 2011, pp. 42–48.

7. Los Angeles from Seeclickfix.com; "Safecast" at blog.safecast.org.

8. Amitabh Kant, "A Tale of India's Cities," *Times of India*, 4 August 2011; McKinsey Global Institute, "Preparing for China's Urban Billion," February 2009, at www.mckinsey.com/Insights/MGI/Research/Urbanization/Preparing_for_urban_billion_in_China.

9. William Oei, "Smart+Connected Community Services to Roll-out Shortly in Songdo" (blog), 8 July 2011.

10. Living PlanIT, "PlanIT Valley—The Benchmark for Future Cities and Sustainable Urban Communities," at living-planit.com/planit_valley.htm.

11. Nicolai Ouroussoff, "In Arabian Desert, A Sustainable City Rises," *New York Times*, 26 September 2010; Dr Sultan Ahmed Al Jaber, "We Won IRENA—Now Let's Prove the Critics Wrong," Masdar, at www.masdar.ae/en/CEO/Desc.aspx?CEO_ID=7&MenuID=55&CatID=77&mnu=cat.

12. Jeremy Kahn, "India Invents a City," *Atlantic Monthly*, July/August 2011; Box 5–1 from "What Is CNU?" Congress for the New Urbanism, at www.cnu.org/who_we_are.

13. GE, "Port of Rotterdam Sailing to Sustainability on Tech Wave," at www.gereports.com/port-of-rotterdam-sailing-to-sustainability-on-tech-wave.

14. Institute for the Future, *A Planet of Civic Laboratories: The Future of Cities, Information and Inclusion* (Palo Alto, CA: 2010); GE, "What Is a Sustainable City?" at www.ge-cities.com/en_GB/Sustainable-Cities.

15. IBM, "IBM Opens $50 Million Smarter Cities Challenge Grant Program to 2012 Applicants," press release (Dubai: 24 October 2011).

16. "London Unveils Digital Datastore," *BBC News*, 7 January 2010.

17. Graduate School of Architecture, Planning and Preservation, Columbia University, "Spatial Information Design Lab Projects," at www.spatialinformationdesignlab.org/projects.php?id=16.

18. SENSEable City Lab, Live Singapore! "The Real-Time City Is Now Real," at senseable.mit.edu/livesingapore.

19. "NYC Big Apps 2.0," at 2010.nycbigapps.com/submissions; "Apps for Democracy," at www.appsfordemocracy.org.

20. "Ushahidi," at www.ushahidi.com/about-us; Social Development Network, "Millions Lost at Water Ministry," 28 April 2010, at www.sodnet.org; Transparency International, *Corruptions Perception Index 2011* (Berlin: 2011).

21. Glover Wright et al., *Report on Open Government Data in India* (Bangalore, India: Centre for Internet and Society, undated), p. 4

22. "The International Aid Transparency Initiative," at www.aidtransparency.net.

23. "Map Kibera" at mapkibera.org.

24. The Public Laboratory for Open Technology and Science, at publiclaboratory.org; West African NGO Network, op. cit. note 1.

25. MySociety.org; "FixMyStreet," at fixmystreet.com; "SeeClickFix," at Seeclickfix.com.

26. Next Drop, op. cit. note 1.

27. "Huduma: Fix My Constituency," at www.dev.huduma.or.ke.

Chapter 6. Measuring U.S. Sustainable Development

1. "Mayor Bloomberg Presents an Update to PLANYC, A Greener, Greater New York," press release (New York: Office of the Mayor, 21 April 2011); Mary Navarro, "City Issues Rule to Ban Dirtiest Oils at Buildings," *New York Times*, 22 April 2011; City of New York, *PlaNYC Update, April 2011* (New York: 2011), p. 179.

2. Mayor's Office of Sustainability, City of Philadelphia, *Executive Summary, Greenworks Philadelphia* (Philadelphia: 2009).

3. U.S. Census, *Statistical Abstract of the United States 2012* (Washington, DC: U.S. Government Printing Office, 2011).

4. T. Hak, B. Moldan, and A. Dahl, eds., *Sustainability Indicators: A Scientific Assessment* (Washington, DC: Island Press, 2007), p. 2; World Commission on Environment and Development, *Our Common Future* (Oxford: Oxford University Press, 1987), p. 43.

5. United Nations, *Agenda 21*, Chapter 40, at www.un.org/esa/dsd/agenda21/res_agenda21_40.shtml.

6. U.N. Conference on Trade and Development, *The Road to Rio* (Geneva: 2011); A. Evans and D. Steven, *Making Rio 2012 Work, Setting the Stage for Global Economic, Social and Ecological Renewal* (New York: Center for International Cooperation, New York University, 2011).

7. United Nations, *Indicators of Sustainable Development: Guidelines and Methodologies, Third Edition* (New York: Division for Sustainable Development, 2007), p. 3.

8. Simon Bell and Stephen Morse, *Sustainability Indicators—Measuring the Immeasurable*, 2nd ed. (London: Earthscan, 2008); Hak, Moldan, and Dahl, op. cit. note 4; Millennium Development Goals, at www.undp.org/mdg/index.shtml.

9. U.S. Office of Management and Budget, *Fiscal Year Analytical Perspectives, Budget of the U.S. Government* (Washington, DC: U.S. Government Printing Office, 2010), pp. 95–101.

10. System of National Accounts, at unstats.un.org/unsd/national account.

11. ICLEI Local Governments for Sustainability USA, at www.icleiusa.org.

12. "Green City Index," Siemens AG, at www.siemens.com/entry/cc/en/greencityindex.htm.

13. Li-Yin Shen et al., "The Application of Urban Sustainability Indicators: A Comparison between Various Practices," *Habitat International*, January 2011, pp. 17–29.

14. ICLEI–USA, *U.S. Local Sustainability Plans and Climate Action Plans* (Boston: 2009); M. Epstein, *Making Sustainability Work, Best Practices in Managing and Measuring Corporate, Social, Environmental and Economic Impacts* (San Francisco: Berrett-Koehler Publishers, Inc., 2008); A. Lynch et al., "Sustainable Development Indicators for the United States," *Penn IUR White Paper in Sustainable Urban Development* (Philadelphia: University of Pennsylvania, Penn Institute for Urban Research, 2011); City of New York, op. cit. note 1.

15. Box 6–1 from Partnership for Sustainable Communities, "HUD, DOT and EPA Partnership:

Sustainable Communities, June 16, 2009," at www.sustainablecommunities.gov.

16. Box 6–2 from ibid.

17. In addition to a new site for the entire effort (www.sustainablecommunities.gov), the Department of Transportation created a dedicated section of its own website at www.dot.gov/livability.

18. "Sustainable Housing and Communities," U.S. Department of Housing and Urban Development (HUD), at portal.hud.gov/hudportal/HUD?src=/program_offices/sustainable_housing_communities.

19. U.S. Department of Transportation, "DOT Livability," at www.dot.gov/livability; U.S. Environmental Protection Agency, "HUD-DOT-EPA Partnership for Sustainable Communities," at www.epa.gov/smartgrowth/partnership/index.html.

20. U.S. Census, op. cit. note 3; LaHood quote from "About Us," Partnership for Sustainable Communities, at www.sustainablecommunities.gov.

21. E. Birch S. and Wachter, *Growing Greener Cities, Urban Sustainability in the 21st Century* (Philadelphia: University of Pennsylvania Press, 2006); M. Kahn, *Green Cities: Urban Growth and the Environment* (Washington, DC: Brookings Institution Press, 2006); R. Ewing and R. Cervero, "Travel and the Built Environment—A Meta-Analysis," *Journal of the American Planning Association*, vol. 76, no. 3 (2010), pp. 265–94; M. Boarnet et al., "The Street Level Built Environment and Physical Activity and Walking: Results of a Predictive Validity Study for the Irvine Minnesota Inventory," *Environment and Behavior* (forthcoming).

22. HUD, "HUD Awards $2.5 Million for Sustainable Communities Research Grant Program," press release (Washington, DC: 16 September 2011).

23. U.S. Department of Transportation, *Office of Management and Budget, Energy & Sustainability Efforts at U.S. DOT*, March 2011, at www.dot.gov/docs/dot_scorecard.pdf.

24. Figure 6–1 is from S. Andreason et al., "Pre-

sentation to the Sustainable Urban Development Working Group," Washington, DC, November 2010.

25. Lynch et al., op. cit. note 14.

26. Ibid.

27. Shen et al., op. cit. note 13.

28. Table 6–1 is from Lynch et al., op. cit. note 14.

29. Bloomberg quote from "Michael Bloomberg Delivers PlaNYC: A Greater, Greener New York," press release (New York: Office of the Mayor, 22 April 2007); Secretary Donovan quote from *Partnership for Sustainable Communities: A Year of Progress for American Communities* (Washington, DC: U.S. Environmental Protection Agency, 2010).

Chapter 7. Reinventing the Corporation

1. Ban Ki-moon, Keynote Address, Redefining Sustainable Development Panel, World Economic Forum Annual Meeting, Davos, 28 January 2011.

2. Michael Narberhaus, "Civil Society Organizations: Time for Systemic Strategies," Great Transition Initiative (GTI), Boston, October 2011; Paul Raskin, "Imagine All the People: Advancing a Global Citizens Movement," GTI, Boston, December 2010.

3. Box 7–1 from Allen L. White, *Transforming the Corporation*, GTI Paper Series No. 5 (Boston: GTI, 2006).

4. U.N. Conference on Trade and Development (UNCTAD), *UNCTAD Training Manual on Statistics for FDI and the Operations of TNCs: Vol. II, Statistics on the Operations of Transnational Corporations* (Geneva: 2009).

5. Figure 7–1 from UNCTAD, *World Investment Prospects Survey 2009–2011* (New York: 2009), p. 31.

6. Aaron Cramer and Zachary Karabell, *Sustainable Excellence: The Future of Business in a Fast Changing World* (New York: Rodale Press, 2010).

7. "The World's Biggest Companies," *Forbes*, at www.forbes.com/global2000/list.

8. Allen L. White, *A New Social Contract: Rethinking Business-Society Relations in the 21st Century*, BSR Occasional Paper (BSR, 2007).

9. Steve Waddell, *Global Action Networks: Creating the Future Together* (New York: McMillan Palgrave, 2010).

10. Figure 7–2 and general information from Global Compact website, at www.unglobalcompact.org.

11. For criticism, see Papa Louis Fall and Mohamed Mounir Zahran, *United Nations Corporate Partnerships: The Role and Functioning of the Global Compact* (Geneva: Joint Inspection Unit, United Nations, 2010).

12. Figure 7–3 from Global Reporting Initiative (GRI) website, at www.globalreporting.org. Allen White is cofounder and former CEO of GRI.

13. Magnus Frostenson, Karolina Windell, and Tommy Borglund, "Mandatory Sustainability Reporting in Swedish State-Owned Companies, Perspectives and Consequences," Department of Business Studies, Uppsala University, undated; German Council for Sustainable Development, "The German Sustainability Code," October 2011; "King Report on Corporate Governance," Wikipedia, undated.

14. Allen L. White, "Why We Need Global Standards for Corporate Disclosure," *Law and Contemporary Problems*, summer 2006.

15. Robert G. Eccles and Michael P. Krzus, *One Report: Integrated Reporting for a Sustainable Strategy* (Hoboken, NJ: John Wiley & Sons, 2010); "Integrated Reporting," at www.theiirc.org.

16. "What Is the Global Compact?" at www.unglobalcompact.org; GRI, "Sustainability Disclosure Database," at database.globalreporting.org; Social Accountability International, *Human Rights at Work: 2010 Annual Report* (New York: 2010).

17. Box 7–2 from the following: Donella H. Meadows, "Envisioning a Sustainable World," presented at Third Biennial Meeting of the International Society for Ecological Economics, San Jose, Costa Rica, 24–28 October 1994, p. 1; Adaptive Edge, *Detroit Wayne County Health Authority*, 2012, at www.adaptive-edge.com; Pieter le Roux et al., "The Mont Fleur Scenarios: What Will South Africa Be Like in the Year 2002?" *Deeper News*, vol. 7, no. 1, pp. 1–22; Dinokeng Scenario Team, *The Dinokeng Scenarios: 3 Futures for South Africa*, 2009, at dinokengscenarios.co.za; Cisco Systems and Global Business Network (GBN), *The Evolving Internet: Driving Forces, Uncertainties and Four Scenarios to 2025* (San Francisco: GBN, 2010); World Business Council for Sustainable Development, *Vision 2050: The New Agenda for Business* (Geneva: 2010); GTI, at www.gtinitiative.org.

18. Allen L. White, *Back to the Future of CSR*, BSR Occasional Paper (BSR, 2011).

19. Allen L. White, "Principles of Corporate Redesign," Center for Progressive Reform workshop, University of North Carolina Law School, 2007.

20. See New Economy Network, at www.neweconomynetwork.org, and "Premises for a New Economy: An Agenda for Rio+20," report of workshop organized by U.N. Division for Sustainable Development, New York, 8–10 May 2010.

21. Corporation 20/20, at www.corporation2020.org.

22. Marjorie Kelly, "Not Just for Profit: Emerging Alternatives to the Shareholder–Centric Model," *Strategy and Business*, spring 2009.

23. "B Corp Legislation," at www.bcorporation.net/publicpolicy; "Governor Brown Signs Legislation to Spur Creation of High Quality Jobs; U.S.'s Largest Economy Accelerates National Benefit Corporation Movement," *CSR Wire*, 10 October 2011.

24. Sheila Shayon, "California Law Creates New 'Flexible Purpose' Category of Positive Impact Corporation," Brand Channel, 17 October 2011.

25. Allen L. White, "When the World Rules Corporations: Pathway to A Global Corporate Charter," GTI, Boston, August 2010.

26. Gar Alperovitz, "Worker-Owners of America, Unite!" (op ed), *New York Times*, 14 December 2011; Marjorie Kelly and Allen White, "Corporate Design: The Missing Business and Public Policy Issue of Our Time," Tellus Institute, Boston, November 2007; Nicholas G. Luviene, "Mondragon: The History of a Movement," posted at Jeffrey Hollender Partners, at www.jeffreyhollender .com; Italy from Italian National Institute of Statistics, at www.istat.it, 2008, and from Unioncamere Lombardia, at www.lom.camcom.it/browse.asp ?goto=1594&livello=0; John Lewis Partnership from Kelly, op. cit. note 22, p. 56.

27. Kelly, op. cit. note 22.

28. Forum for Sustainable and Responsible Investment, *Socially Responsible Investing Trends in the United States* (Washington, DC: 2010).

29. KPMG Global Sustainability Services and U.N. Environment Programme, *Carrots and Sticks for Starters* (Parktown, South Africa, and Nairobi: 2010).

30. Global Impact Investing Network, at www.thegiin.org/cgi-bin/iowa/home/index.html; Global Initiative for Sustainability Ratings, at www.ratesustainability.org.

31. Manuel Escudero and Gavin Power, *Moving Upwards: The Involvement of Boards of Directors in the UN Global Compact* (New York: UN Global Compact, March 2010).

32. Allen L. White, "The Boardroom Imperative: Redefining Corporate Governance in the 21st Century," Keynote Address, UN Global Compact–US Network, San Francisco, 19 October 2009.

33. Ban quoted in Ruth Currant and Alice Chapple, *Overcoming the Barriers to Long-term Thinking in Financial Markets* (Dorking, U.K.: Friends Provident Foundation and Forum for the Future, 2011).

34. Brazilian Institute of Corporate Governance, at www.ibgc.org.br/Secao.aspx?CodSecao=89.

35. Charles Handy, "What's a Business For?" *Harvard Business Review*, December 2002, pp. 3–8.

36. "The Remedies for Capitalism – Paul Polman CEO Unilever," Business in the Community, 4 June 2011, at www.bitc.org.uk/media_centre/ comment/the_remedies_for.html210.

Chapter 8. A New Global Architecture for Sustainability Governance

1. See, for example, the evolution of the U.S. position, at United States "U.S. input to the Belgrade Process," 8 August 2009, and United States, "Sustainable Development for the Next Twenty Years," Input for the 2012 UNCSD Rio + 20 compilation document, 1 November 2011; see also European Union, "Contribution of the European Union and its Member States to the UN Department of Economic and Social Affairs," Input for the 2012 UNCSD Rio + 20 compilation document, 1 November 2011.

2. Kate O'Neill, *From Stockholm to Johannesburg and Beyond: The Evolving Meta-Regime for Global Environmental Governance*, presented at the 2007 Amsterdam Conference on the Human Dimensions of Global Environmental Change, 24–26 May 2007.

3. Jacques Chirac, *Statement of The French Republic to the World Summit on Sustainable Development*, Johannesburg, South Africa, 2 September 2002.

4. U.N. Environment Programme (UNEP), "Elaboration of Ideas for Broader Reform of International Environmental Governance," Co-Chairs of the Consultative Group, 27 October 2010; Second Meeting of the Consultative Group of Ministers or High-level Representatives on International Environmental Governance, Helsinki, 21–23 November 2010; Nairobi-Helsinki Outcome, "Consultative Group of Ministers or High-level Representatives on International Environmental Governance," UNEP, 23 November 2010.

5. Box 8–1 from Nairobi-Helsinki Outcome, op. cit. note 4.

6. Enrique Berruga and Peter Maurer, "Informal Consultations on Environmental Activities Co-Chairs Summary" in Lydia Swart and Estelle Perry, eds., *Global Environmental Governance: Perspectives on the Current Debate* (New York: Center for UN Reform Education, 2007), pp. 16–25; see also UNEP, op. cit. note 4.

7. Richard N. Gardner, "The Role of the U.N. in Environmental Problems" *International Organization*, spring 1972, pp. 237–54.

8. Richard N. Gardner, "U.N. as Policeman," *Saturday Review*, 7 August 1971, p. 47; John W. McDonald, *The Shifting Grounds of Conflict and Peacebuilding: Stories and Lessons* (Lanham, MD: Lexington Books, 2008); Gardner, op. cit. note 7; U.N. General Assembly, "Resolution 2997 (XXVII): Institutional and Financial Arrangements for International Environmental Cooperation," 15 December 1972. The Environmental Coordination Board merged with the U.N. Administrative Committee on Coordination in 1978.

9. Gardner, op. cit. note 7; David Wightman, "Alternative Institutional Arrangements," prepared for the Secretariat of the 1972 Stockholm Conference, unpublished, undated.

10. George F. Kennan, "To Prevent a World Wasteland: A Proposal," *Foreign Affairs*, April 1970, pp. 401–13; Gardner, op. cit. note 7.

11. Gardner, op. cit. note 7.

12. Ibid., pp. 240–41.

13. Additional models are available if an organization is outside the U.N. system. For example, the World Trade Organization, the International Energy Agency, the International Renewable Energy Agency, and the Organisation for Economic Co-operation and Development are all outside the U.N. system and have different governance structures.

14. Examples of subsidiary bodies include the U.N. Development Programme, UNICEF, the U.N. Institute for Training and Research, the Human Rights Council, and the recently created UN Women.

15. Examples of the agencies include the Food and Agriculture Organization, the International Labour Organization, the World Health Organization, and UNESCO.

16. Wightman, op. cit. note 9.

17. U.N. General Assembly, "Preparatory Committee for the United Nations Conference on the Human Environment—Report of the Preparatory Committee on Its Third Session," 30 September 1971, p. 8.

18. Secretary of State's Advisory Committee on the 1972 United Nations Conference on the Human Environment, *Stockholm and Beyond: Report* (Washington, DC: Government Printing Office, 1972), p. 25. Box 8–2 from Maurice F. Strong, "Development, Environment and the New Global Imperatives: The Future of International Co-operation," speech at Carlton University, Ottawa, 1971.

19. Wightman, op.cit. note 9.

20. Gordon Harrison, "Is There a United Nations Environment Programme? Special Investigation at the Request of the Ford Foundation," unpublished, 1997.

21. Michael N. Barnett and Martha Finnemore, *Rules for the World: International Organizations in Global Politics* (Ithaca, NY: Cornell University Press, 2009).

22. Harrison, op. cit. note 20, p. 38.

23. R. Castro and B. Hammond, "The Architecture of Aid for the Environment: A Ten Year Statistical Perspective," *Concessional Finance and Global Partnerships Vice Presidency Working Paper Series No. 3* (Washington, DC: World Bank, 2009).

24. Frank Biermann, "Reforming Global Environmental Governance: The Case for a United Nations Environment Organisation," *Think Piece Series for the 2012 U.N. Conference on Sustainable Development, Stakeholder Forum*, February 2011.

25. Data and Figure 8–1 from Maria Ivanova, *Financing Environmental Governance: Lessons from the United Nations Environment Programme*, Governance and Sustainability Issue Brief Series (Boston: Center for Governance and Sustainability, University of Massachusetts, 2011).

26. Maria Ivanona, "UNEP in Global Environmental Governance: Design, Leadership, Location," *Global Environmental Politics*, February 2010, pp. 30–59.

27. Division of Communications and Public Information, UNEP, at UNEP.org/DCPI; Khatchig Mouradian, "Time in the Wilderness: UNEP in the World Public Consciousness," prepared for graduate course in International Organizations and Environmental Governance, McCormack Graduate School of Policy and Global Studies, University of Massachusetts Boston, fall 2011.

28. U.N. General Assembly, op. cit. note 8.

29. Peter B. Stone, *Did We Save the Earth at Stockholm?* (London: Earth Island, 1973), p. 132.

30. For a complementary and more detailed proposal for reform, see John Scanlon, "Enhancing Environmental Governance for Sustainable Development: Some Personal Reflections," submitted to the preparatory process for the World Congress on Justice, Governance and Law for Environmental Sustainability, October 2011.

Chapter 9. Nine Population Strategies to Stop Short of 9 Billion

1. Taken from the 2010 medium population projection of the U.N. Population Division, available at esa.un.org/unpd/wpp/Excel-Data/population.htm, viewed 2 November 2011; John Bongaarts and Rodolfo A. Bulatao, eds., *Beyond 6 Billion: Forecasting the World's Population* (Washington, DC: National Academy Press, 2000); Wolfgang Lutz, Warren Sanderson, and Sergei Scherbov, "Probabilistic Population Projections Based on Expert Opinion," in Wolfgang Lutz, ed., *The Future Population of the World* (London: Earthscan, 1998).

2. Figures 9–1 and 9–2 from U.N. Population Division, op. cit. note 1.

3. Data on contraceptive prevalence and average

family size from U.N. Population Division, op. cit. note 1; Sushila Singh et al., *Adding It Up: The Costs and Benefits of Investing in Family Planning and Maternal and Newborn Health* (New York: Guttmacher Institute: 2009); demographic evidence from Robert Engelman, "An End to Population Growth: Why Family Planning Is Key to a Sustainable Future," *Solutions*, April 2011.

4. Guttmacher Institute, *In Brief: Facts on Investing in Family Planning and Maternal and Child Health* (New York: 2010); $42 billion for pet food from Erik Assadourian, "The Rise and Fall of Consumer Cultures," in Worldwatch Institute, *State of the World 2010* (New York: W. W. Norton & Company, 2010), p. 16. Box 9–1 from the following: Robert Vale and Brenda Vale, *Time to Eat the Dog? The Real Guide to Sustainable Living* (London: Thames & Hudson: 2009), pp. 225–53; Global Footprint Network, *The Ecological Footprint Atlas 2008*, rev. ed. (Oakland, CA: 2008); Cuba and Haiti calculation by Erik Assadourian based on Vale and Vale, op cit. this note, and on Global Footprint Network, op .cit. this note; Amanda Lilly, "The True Cost of Owning a Pet," *Kiplinger*, September 2011; Pet Airways website, at www.petairways.com; pet food from Elizabeth Higgins, "Global Growth Trends: Sales in the Premium Segments Are Outpacing the Mid-Priced and Economy Segments," *Petfoodindustry.com*, 21 May 2007; Shanghai from Chris Hogg, "Shanghai Announces 'One-Dog Policy,'" BBC, 24 February 2011, and from Elaine Kurtenbach, "Shanghai's One-Dog Policy Causes Anguish for Some Owners," *Huffington Post*, 14 May 2011; humanization from Packaged Facts Pet Analyst David Lummis, *U.S. Pet Market Outlook 2009–2010: Surviving and Thriving in Challenging Times*, PowerPoint presentation, at www.packagedfacts.com/Pet-Outlook-Surviving-2154192/; Kimberly Garrison, "Pet Owners Should Get Fat Cats and Dogs in Shape," *Philly.com*, 17 March 2011.

5. Martha Campbell, Nuriye Nalan Sahin-Hodoglugil, and Malcolm Potts, "Barriers to Fertility Regulation: A Review of the Literature," *Studies in Family Planning*, June 2006, pp. 87–98; Americans' support for access to contraception from National Family Planning & Reproductive Health Association, "Family Planning Facts: Poll Finds Support for Access to Contraception," at www.nfprha.org/main/family_planning.cfm?Catego ry=Public_Support&Section=Access_Poll.

6. Dina Abu-Ghaida and Stephan Klasen, "The Costs of Missing the Millennium Development Goal on Gender Equity," *World Development*, July 2004, pp. 1,075–107.

7. International Institute for Applied Systems Analysis (IIASA), discussion with author, cited in Robert Engelman, "Population & Sustainability: Can We Avoid Limiting the Number of People?" *Scientific American Earth 3.0*, summer 2009, pp. 22–29.

8. Educational attainment estimates for 1970–2000 by Wolfgang Lutz et al., "Reconstruction of Population by Age, Sex and Level of Educational Attainment of 120 Countries for 1970–2000," *Vienna Yearbook of Population Research* (Laxenburg, Austria: IIASA, 2007), pp. 193–235; projections by Samir K. C. et al., "Projection of Populations by Level of Educational Attainment, Age, and Sex for 120 Countries for 2005–2050," *Demographic Research*, vol. 22, no. 15 (2010), pp. 383–472; both datasets extrapolated by author to world population from U.N. Population Division, *World Population Prospects: The 2008 Revision Population Database*, which has since been superseded with some modest changes in population estimates by a 2010 revision (see U.N. Population Division, op. cit. note 1); remaining gender gap from World Bank, *Getting to Equal: Promoting Gender Equality through Human Development* (Washington, DC: 2011).

9. Robert Engelman, "Women Slowly Close Gender Gap With Men," *Vital Signs Online*, 9 March 2011.

10. Ulla Larson and Marida Hollos, "Women's Empowerment and Fertility Decline among the Pare of Kilimanjaro Region, Northern Tanzania," *Social Science & Medicine*, vol. 27, pp. 1,099–115.

11. These surveys are available at www.measuredhs.com, viewed 8 November 2011.

12. Trisha E. Mueller, Lorrie E. Gavin, and Aniket Kulkarni, "The Association between Sex Education and Youth's Engagement in Sexual Intercourse, Age at First Intercourse, and Birth Control Use at First Sex," *Journal of Adolescent Health*, January 2008, pp. 89–96.

13. Lack of impacts from pro-natalist payment for births from "Eliminating Targets, Incentives, and Disincentives," in Population Information Program, *Informed Choice in Family Planning: Helping People Decide, Population Reports*, spring 2001; baby bonuses in Russia from Daniel Gross, "Children for Sale: Would $36,000 Convince You to Have Another Kid?" *Slate.org*, 24 May 2006; baby bonuses in Singapore from Government of Singapore, "Child Development Credits," undated, at www.babybonus.gov.sg/bbss/html/index.html.

14. Noriko O. Tsuyo, "Fertility and Family Policies in Nordic Countries, 1960–2000," *Journal of Population and Social Security (Population)*, Supplement to Volume 1 (Tokyo: National Institute of Population and Social Security Research), at www.ipss.go.jp/webj-ad/WebJournal.files/population/2003_6/4.Tsuya.pdf.

15. Population Connection website, at www .populationeducation.org/index.php?option=com _content&view=article&id=1&Itemid=2.

16. Warren Sanderson and Sergei Scherbov Sanderson, "Remeasuring Aging," *Science*, 10 September 2010, pp. 1,287–88.

Chapter 10. From Light Green to Sustainable Buildings

1. U.N. Environment Programme (UNEP), *Towards a Green Economy: Pathways to Sustainable Development and Poverty Eradication* (Nairobi: 2011).

2. Population Division, *2009 Revision of World Urbanization Prospects* (New York: United Nations, 2010).

3. Table 10–1 from Kaarin Taipale, "Buildings and Construction as Tools for Promoting More Sustainable Patterns of Production and Consumption," *Sustainable Development Innovation Briefs*, U.N. Department of Economic and Social Affairs, New York, March 2010.

4. Charles Kibert, "Progress in the Design and Deployment of Net Zero Energy Buildings in the U.S.," presented at the World Sustainable Building Conference, Helsinki, October 2011.

5. Marrakech Task Force on Sustainable Buildings and Construction, *Buildings for a Better Future* (Helsinki: Ministry of the Environment of Finland, 2007); Husbanken, "Social Housing, Universal Design and Environmentally Friendly Housing," at www.husbanken.no/english/social-housing-universal-design-and-environmental-issues.

6. Sonja Köppel and Diana Ürge-Vorsatz, *Assessment of Policy Instruments for Reducing Greenhouse Gas Emissions from Buildings*, Report for the UNEP SBCI (Budapest: Central European University, 2007).

7. Tatiana de Feraudy, Rajat Gupta, and Niclas Svenningsen, *Sustainable Solutions for Social Housing: Guidelines for Project Developers* (unpublished draft) (Nairobi: UNEP, 2011); Sustainable Building Process, 2010 research project led by Dr. Tarja Häkkinen at VTT Technical Research Centre of Finland.

8. Box 10–1 from Catherine Stansbury and Neill Stansbury, *Examples of Corruption in Infrastructure* (Old Amersham, U.K.: Global Infrastructure Anti-Corruption Centre, 2008); Transparency International, "Preventing Corruption on Construction Project," at www.transparency.org/tools/contracting/construction_projects.

9. Umberto Berardi, "Comparison of Sustainability Rating Systems for Buildings and Evaluation of Trends," presented at the World Sustainable Building Conference, Helsinki, October 2011; Taipale, op. cit. note 3; Box 10–2 from Bruno Mesureur, "Common Metrics for Key Issues: A Proposal for the Sustainable Building Alliance," PowerPoint presentation, at www.sballiance.org/dldocuments/20091105-SBA-core-set-of-indicators.pdf.

10. Aiste Blaviesciunaite and Raymond J. Cole, "The Cultural Values Embedded in Building Environmental Assessment Methods: A Comparison of LEED© and CASBEE," presented at the World Sustainable Building Conference, Helsinki, October 2011; Alsema Erik et al., "Potential of Sustainable Building Assessment Methods as Instruments of Steering of Sustainable Building," Deliverable 3.2 of the *SuPerBuildings* Project of the FP7 Cooperation Programme of the European Commission, 2011.

11. Clearingstelle EEG, at www.clearingstelle-eeg.de/english; "Bundestag Adopts Modification of Solar Power Feed-in Tariffs," press release (Bonn: German Ministry of the Environment, 6 May 2010).

12. Intergovernmental Panel on Climate Change, *Climate Change 2007: Fourth Assessment Report* (Cambridge, U.K.: Cambridge University Press, 2007); Box 10–3 from the following: "Energy Efficiency: Energy Performance of Buildings," *Europa: Summaries of EU Legislation*, at europa.eu/legislation_summaries/other/l27042_en.htm; European Commission, "Energy Efficiency of Buildings," at ec.europa.eu/energy/efficiency/buildings/buildings_en.htm; European Commission, Joint Research Centre, "EU Ecolabel and Green Public Procurement for Buildings," at susproc.jrc.ec.europa.eu/buildings/index.html.

Chapter 11. Public Policies on More-Sustainable Consumption

1. Ricardo Barros et al., *A Nova Classe Média Brasileira: Desafios que Representa para a Formulação de Políticas Públicas* (Brazil: Secretaria de Assuntos Estratégicos da Presidência da República, August 2011); "Governo Traça Perfil da Nova Classe Média Brasileira" (blog), *Blog de Planalto*, 8 August 2011.

2. Editora Abril e Data Popular, "As Ponderosas da Nova Classe Média," survey in March–July 2011.

3. Zenith Optimedia, "Quadrennial Events to Help Ad Market Grow in 2012 Despite Economic Troubles," press release (London: 5 December 2011).

4. WWF, ZSL, and GFN, *Living Planet Report 2010* (Gland, Switzerland: 2010), p. 8.

5. Sustainable Europe Research Institute, GLOBAL 2000, and Friends of the Earth Europe,

Overconsumption? Our Use of the World's Natural Resources (September 2009).

6.   Erik Assadourian, "The Rise and Fall of Consumer Cultures," in Worldwatch Institute, *State of the World 2010* (New York: W. W. Norton & Company, 2010), p. 4.

7.   Ibid., p. 6. Box 11–1 from the following: Elizabeth W. Dunn, Daniel T. Gilbert, and Timothy D. Wilson, "If Money Doesn't Make You Happy, Then You Probably Aren't Spending It Right," *Journal of Consumer Psychology*, vol. 21 (2011), pp. 115–25; Gary Gardner and Erik Assadourian, "Rethinking the Good Life," in Worldwatch Institute, *State of the World 2004* (New York: W. W. Norton & Company, 2004), pp. 164–79; Robert D. Putnam, *Bowling Alone: The Collapse and Revival of American Community* (New York: Simon & Schuster, 2000), N. Marks et al., *The (un)Happy Planet Index: An Index of Human Well-being and Ecological Impact* (London: NEF, 2009), p. 28; Burlington study from Kenneth Mulder, Robert Costanza, and Jon Erickson, "The Contribution of Built, Human, Social and Natural Capital to Quality of Life in Intentional and Unintentional Communities," *Ecological Economics*, August 2006, pp. 13–23; "Tool Library Welcomes Borrowers in Columbus," *Columbus Dispatch*, 19 October 2009; Lucie K. Ozanne and Julie L. Ozanne, "Parental Mediation of the Market's Influence on their Children: Toy Libraries as Safe Havens," paper presented at The Academy of Marketing Conference, Leeds, U.K., 7–9 July 2009; Laura Stoll, "Beyond GDP: UK To Measure Well-Being," *Yes Magazine*, 27 January 2011; Welsh Government Official Website, "Wales Ecological Footprint–Scenerios to 2020," 27 May 2008; Aaron Best et al., "Potential of the Ecological Footprint for Monitoring Environmental Impacts from Natural Resource Use: Analysis of the Potential of the Ecological Footprint and Related Assessment Tools for Use in the EU's Thematic Strategy on the Sustainable Use of Natural Resources," Report to the European Commission, 2008.

8.   Nielsen, "Sustainability Survey: Global Warming Cools Off as Top Concern," press release (New York: 28 August 2011); World Economic Forum, *Sustainability for Tomorrow's Consumer, The Business Case For Sustainability* (Geneva: 2009); World

Economic Forum, *Redesigning Business Value, A Roadmap for Sustainable Consumption* (Geneva: 2010); World Economic Forum, *The Consumption Dilemma, Leverage Points to Accelerate Sustainable Growth* (Geneva: 2011).

9.   Rainforest Action Network, *Greatest Hits, 1985–2010* (San Francisco: 2010), p. 6; Greenpeace International, "Sweet Success for Kit Kat Campaign: You Asked, Nestlé Has Answered," Feature Story, 17 May 2010.

10.   GlobeScan, "Social Media Users 'More Active' As Ethical Consumers: Global Poll," press release (London: 20 July 2011).

11.   Ibid.

12.   "Arezzo Desiste de Usar Pele de Raposa e de Coelho em sua Nova Coleção," *O Estado de São Paulo*, 18 April 2011.

13.   Daniela Siaulys et al., "Estudo: A Sustentabilidade na Visão do Consumidor Global," GS&MD – Gouvêa de Souza, October 2010; Government of Sweden, "New Green Cars to be Exempted from Vehicle Tax," press release (Stockholm: 10 March 2009), Enda Curran and Ray Brindal, "Australia's Carbon Tax Clears Final Hurdle," *Wall Street Journal*, 8 November 2011.

14.   Akatu Institute, "Videos," at www.akatu.org.br/videos.

15.   Robin Andersen and Pamela Miller, "Media Literacy, Citizenship, and Sustainability," in Worldwatch Institute, op. cit. note 6, pp. 157–63.

16.   Prefeitura de São Paulo, "São Paulo Pode se Tornar uma Cidade Limpa," 7 January 2009; Amy Curtis, "Five Years After Banning Outdoor Ads, Brazil's Largest City Is More Vibrant Than Ever" (blog), The Center for a New American Dream, 8 December 2011.

17.   "Precautionary Purchase Ordinance," California Sustainability Alliance, at sustainca.org/tools. Box 11–2 based on the following: Ministry of Environment, Government of Japan, "The Basic Environment Plan" at www.env.go.jp/en/policy/plan/basic/foreword.html; Ministry of Environment, Government of Japan, *Establishing a Sound Material-Cycle Society* (Tokyo: 2010); DOWA Eco-System Co., Ltd., "Resource Recycling," at www.dowa-eco.co.jp/en/recycle.html; Ministry of Environment, Government of Japan, "Basic Act Establishing a Sound Material Cycle Society," at www.env.go.jp/en/laws/recycle/12.pdf; Ministry of Environment, Government of Japan, *The World in Transition and Japan's Efforts to Establish a Sound Materials-Cycle Society* (Tokyo: 2008); Martin Fackler, "Japan Split on Hope for Vast Radiation Cleanup," *New York Times*, 6 December 2011.

18. Alana Institute, "Empresas Anunciam Restrição de Publicidade de Alimentos para Crianças," 26 August 2009, at www.alana.org.br; "Conar Contra Greenwashing," Ideia Sustentável, 13 June 2011, at www.ideiasustentavel.com.br.

19. Akatu Institute, "Ministério do Meio Ambiente Lança Campanha pela Redução de Sacos Plásticos,"at www.akatu.org.br, 22 June 2009; "MMA: Campanha 'Saco é um Saco' Evita Consumo de 5Bilhões de Sacolas Plásticas," 6 January 2011, at www.ecodebate.com.br; Brandon Mitchener, "Sweden Pushes Its Ban on Children's Ads," *Wall Street Journal*, 29 May 2001.

20. "Top Runner Program: Developing the World's Best Energy-Efficient Appliances," The Energy Conservation Center, Japan, undated.

21. Sheri Todd, *Improving Work-Life Balance—What Are Other Countries Doing?* (Ottawa: Human Resources and Skills Development Canada, 2004).

22. Centre for Bhutan Studies, "Results of the Second Nationwide 2010 Survey on Gross National Happiness," at www.grossnationalhappiness.com; Commission on the Measurement of Economic Performance and Social Progress, at www.stiglitz-sen-fitoussi.fr/en/index.htm.

23. Organisation for Economic Co-operation and Development, "OECD Welcomes Experts' Call on Need for New Measures of Social Progress," press release (Paris: 14 September 2009). Box 11–3 from the following: Millennium Consumption Goals Initiative, "Proposal on Millennium Consumption Goals (MCGs): Input to the Rio+20 Compilation Document," at www.uncsd2012.org; Erik Assadourian,

"Millennium Consumption Goals: An Update" (blog), *Transforming Cultures*, 25 April 2011.

24. Good Guide, at www.goodguide.com; Nielsen, op. cit. note 8.

Chapter 12. Mobilizing the Business Community in Brazil and Beyond

1. Population and economic growth from U.N. Environment Programme, *Keeping Track of Our Changing Environment: From Rio to Rio+20 (1992–2012)* (Nairobi: 2011); Millennium Ecosystem Assessment, *Living Beyond Our Means: Natural Assets and Human Well-Being: Statement from the Board* (Washington, DC: World Resources Institute, 2005), p. 2; Johan Rockström et al., "A Safe Operating Space for Humanity," *Nature*, 24 September 2009, pp. 472–75.

2. Hilary French, *Vanishing Borders* (New York: W.W. Norton & Company, 2000).

3. "Objectives & Themes," Rio+20: United Nations Confernce on Sustainable Development website, at www.uncsd2012.org/rio20/objective andthemes.html.

4. Quote is from Kenneth Rogoff, "Modern Capitalism is Not Under Threat—Yet," *Business Day* (South Africa), 7 December 2011.

5. R. Abramovay, *The Transition to a New Economy* (in Portuguese), draft prepared for Avina Foundation, São Paulo, 2011.

6. Commission on Measurement of Economic Performance and Social Progress, *Summary of Recommendations* (in Portuguese), published by the Industry Federation of Paraná, 2011.

7. James Grubel, "Australia Passes Landmark Carbon Price Laws," *Reuters*, 8 November 2011; Enda Curran and Ray Brindal, "Australia's Carbon Tax Clears Final Hurdle," *Wall Street Journal*, 8 November 2011; "Australia Makes Green Cuts to Fund Flood Relief," *Radio Australia*, 28 January 2011; China from "FACTBOX: China's Carbon Market Plans," *Reuters Africa*, 10 November 2011, from Dinakar Sethuraman, "China Considers Rules for Domestic Carbon Trading Proposal, Official Says," *Bloomberg*, 28 October 2010, and from Li Jing, "Carbon Trading in Pipeline," *China Daily*, 22 July 2010.

8. R. Costanza et al., "The Value of the World's Ecosystem Services and Natural Capital," *Nature*, 15 May 1997, pp. 253–60; see also Chapter 16.

9. Bryan Walsh, "Can Ecuador Trade Oil for Forests?" (blog), *Time*, 11 October 2011.

10. See Global Reporting Initiative, at www.global reporting.org/network/report-or explain/Pages/default.aspx.

11. "Relief," Sustainable Procurement Resource Centre, at www.sustainable-procurement.org/about-us/past-projects/relief; Rachel Biderman et al., *Guia de Compras Públicas Sustentáveis* (ICLEI European Secretariat), pp. 23–24.

12. Biderman et al., op. cit. note 11.

13. Ethical Markets, "2011 Ethicmark® Award Announced," press release (St. Augustine, FL: 3 October 2011); "About," Girl Effect website, at girl effect.org/media/about; Hazel Henderson, "Un-Ethical Neuromarketing," *CSR Wire Talk Back*, 14 February 2010; Maria O. Pinochet, "The Current State of Ethical Advertising," Ethicalmarkets.com, 29 December 2011.

14. "Media Literacy Worldwide," European Commission, at ec.europa.eu/culture/media/literacy/global/index_en.htm; Robin Andersen and Pamela Miller, "Media Literacy, Citizenship, and Sustainability," in Worldwatch Institute, *State of the World 2010* (New York: W. W. Norton & Company, 2010), pp. 157–63; David W. Orr, "What Is Higher Education for Now," ibid., pp. 75–82.

15. Rede Nossa São Paulo, Rede Social Brasileira por Cidades Justas e Sustantáveis, and Instituto Ethos de Empresas e Responsabilidade Social, *Sustainable Cities Programme*, at www.cidadessustent avies.org.

16. Steven Greenhouse and Graham Bowley, "Tiny Tax on Financial Trades Gains Advocates," *New York Times*, 6 December 2011.

17. BOVESPA website, at www.bmfbovespa.com.br; 2012 Index Portfolio, at www.bmfbovespa.com.br/Indices/download/Release-Carteira-ISE-2012-Ingles.pdf; Michel Doucin, *Corporate Social Responsibility: Private Self-Regulation is Not Enough*, Private Sector Opinion No. 24 (Washington, DC: Global Corporate Governance Forum, 2011).

18. For the Ethos Indicators Virtual Tool in Box 12–1, see www.ethos.org.br/docs/conceitos_pra ticas/indicadores/default.asp.

19. Climate Forum, at www.forumempresarial peloclima.org.br.

20. "Empresa Pró-Ética," Controladoria Geral da União, at www.cgu.gov.br/empresaproetica.

21. See "Companies and Human Rights in Perspective—Decent Work: Terms of Reference," 2011 (in Portuguese), at www1.ethos.org.br/Ethos Web/arquivo/0-A-cb3MarcoDeReferenciaCOM PLETO.pdf.

22. See Ethos Conference website, at www .ethos.org.br/ce201; see also interview by Ethos vice president Paulo Itacarambi, at www1 .ethos.org.br/EthosWeb/pt/5735/servicos_do_ portal/noticias/itens/conferencia_ethos_2011_ debatera_nova_economia_.aspx (both in Portuguese).

23. Ethos Institute, *Protagonists of a New Economy: Towards Rio+20* (São Paulo: August 2009) (in Portuguese); the conference 2011 was covered by mainstream media, especially by *Jornal Valor Económico*, which dedicated eight pages in its edition of 12 August 2011.

24. For origins of Global Union for Sustainability, see worldforumforsustentability.org.

25. French, op. cit. note 2.

Chapter 13. Growing a Sustainable Future

1. World Bank, *World Development Report 2007* (Washington, DC: 2007).

2. Enough food from U.N. Food and Agriculture Organization (FAO), "Feeding the World, Eradicating Hunger," Background Paper, World Food Summit, Rome, 16–18 November 2009.

3. Undernourished from FAO, *The State of Food Insecurity in the World 2011* (Rome: 2011).

4. Resource degradation from Foresight, *The Future of Food and Farming, Executive Summary* (London: The Government Office for Science, 2011); competing demands from World Food Programme (WFP), "The New Paradigm of Hunger" (blog), Rome, 22 February 2011; freshwater use from Comprehensive Assessment of Water Management in Agriculture, *Water for Food, Water for Life* (London and Colombo, Sri Lanka: Earthscan and International Water Management Institute, 2007); pollution from Bridget R. Scanlon et al., "Global Impacts of Conversions from Natural to Agricultural Ecosystems on Water Resources: Quantity versus Quality," *Water Resources Research*, vol. 43, no. W03437 (2007); declining yields from "Policy Issues in Irrigated Agriculture," in FAO, *The State of Food and Agriculture 1993* (Rome: 1993).

5. Loss of plant genetic resources from FAO, *Save and Grow: A Policymaker's Guide to the Sustainable Intensification of Smallholder Crop Production* (Rome: 2011); species cultivated from The Development Fund/Utviklingsfondet, *A Viable Food Future* (Oslo, Norway: 2010); soil degradation from Foresight, op. cit. note 4.

6. Bertram Zagema, *Land and Power: The Growing Scandal Surrounding the New Wave of Investments in Land*, Briefing Paper (Oxford: Oxfam International, 2011).

7. Development assistance for agriculture, including forestry and fishing, calculated from Organisation for Economic Co-operation and Development (OECD), "Official Bilateral Commitments by Sector," DAC5 database; support for developing-country agriculture from OECD, "Producer Support Estimate by Country," at www.oecd.org/dataoecd/30/58/45560148.xls?contentId=45560149.

8. Monsanto, "Alliance for Abundant Food and Energy to Highlight Promise of Agriculture to Sustainably Meet Food and Energy Needs," press release (Washington, DC: 24 July 2008); Doug Cameron, "Agribusiness Group Forms to Protect Ethanol Subsidies," *Wall Street Journal*, 25 July 2008.

9. Ecological approach required from Amir Kassam et al., "Production Systems for Sustainable Intensification: Integrating Productivity with Ecosystem Services," in *Technikfolgenabschätzung – Theorie und Praxis*, July 2011, pp. 38–45; studies on success of agroecological farming include Beverly D. McIntyre et al., *Agriculture at a Crossroads* (Washington, DC: International Assessment of Agricultural Knowledge, Science and Technology for Development, 2009), Olivier de Schutter, "Agroecology and the Right to Food," U.N. General Assembly, 17 December, 2010, and Hans Herren, "Agriculture: Investing in Natural Capital," in U.N. Environment Programme, *Towards a Green Economy: Pathways to Sustainable Development and Poverty Eradication* (Nairobi: 2011); rice intensification from Africare, Oxfam America, and WWF–ICRISAT, *More Rice for People, More Water for the Planet* (Hyderabad, India: 2010).

10. Investment needed from Josef Schmidhuber and Jelle Bruinsma, "Investing Towards a World Free of Hunger: Lowering Vulnerability and Enhancing Resilience," in Adam Prakash, ed., *Safeguarding Food Security in Volatile Global Markets* (Rome: FAO, 2011).

11. Hungry people in rural areas from Pedro Sanchez et al., *Halving Hunger: It Can Be Done* (New York: Millennium Project, U.N. Development Programme, 2005); yield increases from Africare, Oxfam America, WWF-ICRISAT, op. cit. note 9.

12. Supaporn Anuchiracheeva and Tul Pinkaew, *Jasmine Rice in the Weeping Plain: Adapting Rice Farming to Climate Change in Northeast Thailand* (Oxford: Oxfam GB, 2009).

13. Growth spark from Arabella Fraser, *Harnessing Agriculture for Development*, Research Report (Oxford: Oxfam International, 2009); growth effect on poorest from FAO, *How to Feed the World in 2050* (Rome: 2009); evidence from wealthy countries from Ha-Joon Chang, "Rethinking Public

Policy in Agriculture: Lessons from History, Distant and Recent," *Journal of Peasant Studies*, vol. 36, no. 3 (2009), pp. 477–515.

14. Gender lens from "Women in Agriculture: Closing the Gender Gap for Development," in FAO, *The State of Food and Agriculture 2011* (Rome: 2011); African women's activities from Women in Development Service, *Women and Sustainable Food Security*, Towards Sustainable Food Security Series (Rome: FAO, 1996); systematic exclusion from Agnes R. Quisumbing and Lauren Pandolfelli, *Promising Approaches to Address the Needs of Poor Female Farmers*, Discussion Paper (Washington, DC: International Food Policy Research Institute (IFPRI), 2009); share of aid to women from OECD, *Aid in Support of Gender Equality and Women's Empowerment* (Paris: OECD–DAC Secretariat, 2007); farm output from "Women in Agriculture," op. cit. this note; impact of women's control over income from IFPRI, *Women: The Key to Food Security* (Washington, DC: 2000).

15. Claudia Canepa, "Women-led Dairy Development in Vavuniya, Sri Lanka," at growsellthrive .org/page/dairy-sri-lanka.

16. Sally Baden and Carine Pionetti, *Women's Collective Action in Agricultural Markets: Synthesis of Preliminary Findings from Ethiopia, Mali and Tanzania* (Oxford: Oxfam GB, 2011).

17. Institution of Civil Engineers, Oxfam GB, and WaterAid, *Managing Water Locally, An Essential Dimension of Community Water Development* (London and Oxford: 2011).

18. Oxfam International, *Farmer-to-Farmer Training: A Learning Summary from the Honduras Agricultural Scale Up Programme* (Oxford: Oxfam International, 2009).

19. Decline in agricultural extension from L. van Crowder, *Agricultural Extension for Sustainable Development* (Rome: FAO, 1996); gains from extension services from Chang, op. cit. note 13.

20. Focus of technologies from large companies from Laura German, Jeremias Mowo, and Margaret Kingamkono, "A Methodology for Tracking the 'Fate' of Technological Innovations in Agriculture," *Agriculture and Human Values*, vol. 30, no. 22-16 (2006), pp. 353–69; smallholder innovations ignored from C. Chikozho, "Policy and Institutional Dimensions of Small-holder Farmer Innovations in the Thukela River Basin of South Africa and the Pangani River Basin of Tanzania: A Comparative Perspective," *Physics and Chemistry of the Earth*, 2005, pp. 913–24.

21. Shekhar Anand and Gizachew Sisay, "Engaging Smallholders in Value Chains—Creating New Opportunities for Beekeepers in Ethiopia," in David Wilson, Kirsty Wilson, and Claire Harvey, eds., *Small Farmers Big Change: Scaling Up Impact in Smallholder Agriculture* (Warwickshire and Oxford: Practical Action Publishing Ltd and Oxfam GB, 2011), pp. 53–66.

22. Abdoulaye Dia and Aboubacar Traore, *Effective Cooperation: A New Role for Cotton Producer Co-ops in Mali* (Oxford: Oxfam GB, 2011).

23. Armenia from Arabella Fraser, *Harnessing Agriculture for Development*, Research Report (Oxford: Oxfam International, 2009); Indonesia from internal Oxfam program report, April 2010; Sri Lanka case from Robert Bailey, *Growing a Better Future: Food Justice in a Resource-Constrained World* (Oxford: Oxfam International, 2011).

24. Ethiopia from WFP and Oxfam America, "R4 Rural Resilience Initiative: Partnership for Resilient Livelihoods in a Changing Climate," brochure, Rome and Boston, 2011.

25. Loans and credit for women farmers from FAO, "Agricultural Support System, Gender and Development Plan of Action 2002–2007," Rome, May 2003; savings and credit associations from Emily Alpert, Melinda Smale, and Kelly Hauser, *Investing in Small Farmers Pays: Rethinking How to Invest in Agriculture* (Oxford: Oxfam International, 2009).

26. Donors' pledge from G8 Summit 2009, "'L'Aquila' Joint Statement on Global Food Security," L'Aquila, Italy, 10 July 2009.

Chapter 14. Food Security and Equity in a Climate-Constrained World

1. Jason Straziuso, "KFC Goes to Kenya; First U.S. Fast-Food Chain in E. Africa," *Associated Press*, August 23, 2011.

2. Discussion with author, April 2011.

3. "Exports," National Meat and Poultry Processing Board, Ministry of Food Processing Industries, New Delhi, at nmppb.gov.in/PAGE/Exports.htm.

4. Jesse Chang, "Meat Production and Consumption Contine to Grow," *Vital Signs Online*, 11 October 2011; Food and Agriculture Organization (FAO), *FAOSTAT Statistical Database*, at faostat.fao.org.

5. Chang, op. cit. note 4; FAO, op. cit. note 4; United Nations, Population Division, *World Population Prospects: The 2010 Revision* (New York: 2011).

6. Pawan Kumar, Rabo India Finance Ltd., "Indian Poultry Industry: Overview and Outlook," presented at VIV India 2010, Bangalore, 2010; "Interview: Poultry Consumption Growing Faster in India Compared to Other Meats," *Feedinfo News Service*, 3 January 2010; FAO, op. cit. note 4.

7. Pew Commission on Industrial Farm Animal Production, *Putting Meat on the Table: Industrial Farm Animal Production in America* (Baltimore: Johns Hopkins Bloomberg School of Public Health, 2008), p. 29.

8. Quote and Xu studies from Xiaofeng Guan, "Animal Waste a Heavy Burden for Environment," *China Daily*, 3 March 2007. Box 14–1 from the following: harvest and employment data from FAO, *The State of World Fisheries and Aquaculture 2010* (Rome: 2011) (all 2009 data are provisional estimates); projection for 2020 from Cécile Brugère and Neil Ridler, *Global Aquaculture Outlook in the Next Decades: An Analysis of National Aquaculture Production Forecasts to 2030*, Fisheries Circular No. 1001 (Rome: FAO, 2004); Lake Taihu from L. Cao et al., "Environmental Impact of Aquaculture and Countermeasures to Aquaculture Pollution in China," *Environmental Science and Pollution Research*, vol. 14, no 7 (2007), pp. 452–62.

9. Meng quote from Evan Osnos and Laurie Goering, "World's Giants to Alter Food Equation," *Chicago Tribune*, 11 May 2008; meals and meat in China from "China's Journey," *National Geographic*, May 2008; energy from fats and deaths from chronic diseases from Agriculture and Consumer Protection Department, "Fighting Hunger—and Obesity," *Spotlight*, FAO, 2006; obesity levels from "China: Affluence Brings Obesity," *New York Times*, 7 November 2006.

10. Changes in diets and quote from Frank Hu, "Globalization of Food Patterns and Cardiovascular Disease Risk" (editorial), *Circulation*, 4 November 2008, pp. 1,913–14; President of the 65th Session, "High-Level Meeting on Non-Communicable Diseases," General Assembly of the United Nations, New York.

11. Linda Blake, "Obesity Silent Killer in India," *VOA News*, 30 April 2010; "Latest Diabetes Figures Paint Grim Global Picture," press release (Brussels: International Diabetes Foundation, 18 October 2009); projected rates from K. Srinath Reddy, "Persisting Public Health Challenges," *The Hindu*, 15 August 2007; Economist Intelligence Unit, *The Silent Epidemic: An Economic Study of Diabetes in Developed and Developing Countries* (London: June 2007), p. 3; World Bank, "India At A Glance," 25 February 2011, at devdata.worldbank.org/AAG/ind_aag.pdf.

12. Quote from Jordan Helton, "World Population Reaches 7 Billion," *Global Post*, 30 October 2011.

13. Naylor from Mark Bittman, "Rethinking the Meat Guzzler," *New York Times*, 27 January 2008.

14. U.S. Department of Agriculture (USDA), Foreign Agricultural Service, *Production, Supply and Distribution (PSD) Online*, viewed 22 December 2011; Keith Bradsher, "Rise in China's Pork Prices Signals End to Cheap Output," *New York Times*, 1 June 2007; World Resources Institute, *EarthTrends Searchable Database*, at earthtrends.wri.org.

15. Soya Tech, "Soy Facts," at soyatech.com; China soy purchases from Associação dos Produtores de Soja do Estado de Mato Grosso, *Estatisticas: Exportações*, at www.aprosoja.com.br; China and Brazilian soy from FAO, op. cit. note 4. Box 14–2 from the following: production data from FAO, op. cit. note 4, from Roberto Smeraldi and Peter H. May, *The Cattle Realm: A New Phase in the Livestock Colonization of Brazilian Amazonia*, Highlights in English (São Paulo: Friends of the Earth, Brazilian Amazon, 2008), p. 4, from "In 10 Years Brazil Wants its Market Share in Meat to be 44.5% of World Trade," *BrazzilMag*, 6 March 2010, and from USDA, Economic Research Service, *Briefing Rooms: Agricultural Baseline Projections: Global Agricultural Trade, 2010–2019* (Washington, DC); genetically modified (GM) soy and corn from "Brazil—GM Soy Becomes Market Leader," *Meat Trade News Daily*, 19 December 2009; small farmers and GM crops in Brazil from "Cattle Farming, Soy, and the Increase in Deforestation in the Amazon: Interview with Tatiana de Carvalho," Instituto Humanitas Unisinos, 20 March 2008 (translation from Portuguese by Simone de Lima).

16. Dean Nelson, "India Joins Neocolonial Rush for Africa's Land and Labour," (London) *Telegraph*, 28 June 2009.

17. FAO, *Food Outlook: Global Market Analysis* (Rome: 2007); increase in grain use for feed from FAO, *Food Outlook: Global Market Analysis* (Rome: 2011).

18. Mia MacDonald and Justine Simon, "Climate, Food Security, and Growth: Ethiopia's Complex Relationship with Livestock," Brighter Green, New York, 2011.

19. Henning Steinfeld et al., *Livestock's Long Shadow: Environmental Issues and Options* (Rome: FAO, 2006).

20. "For Want of a Drink: A Special Report on Water," *The Economist*, 22 May 2010; World Water Assessment Programme, *World Water Development Report: Water in a Changing World* (Paris: UNESCO, 2009), p. 36.

21. World Water Assessment Programme, op. cit. note 20, p. 36; quote from Martin Mittelstaedt, "UN Warns of Widespread Water Shortages," (Toronto) *Globe and Mail*, 12 March 2009; beef and water from A. Y. Hoekstra and A. K. Chapagain, "Water Footprints of Nations: Water Use by People as a Function of Their Consumption Patterns," *Water Resources Management*, January 2007, pp. 35–48.

22. M. M. Mekonnen and A. Y. Hoekstra, *The Green, Blue and Grey Water Footprint of Farm Animals and Animal Products, Volume 1: Main Report* (Paris: UNESCO, 2010); "Contrarian Alert, Fishy Jobs Report Details, Getting Water to China...," *Agora Financial*, 9 November 2009.

23. Shama Perveen, "Water: The Hidden Export," *India Together*, August 2004.

24. Steinfeld et al., op. cit. note 19; Robert Goodland and Jeff Anhang, "Livestock and Climate Change," *World Watch*, November/December 2009; FAO, *Environmental Issues and Options* (Rome: 2006).

25. World Wildlife Fund, "Brazil Throws Out Another Climate Challenge Updating Greenhouse Gas Inventory," Washington, DC, 27 November 2009; emissions from agriculture from "Brazil: Economic Structure," *Economy Watch*, 15 March 2010; cattle sector emissions from Mercedes Bustamente et al., *Estimating Recent Greenhouse Gas Emissions from Cattle Raising in Brazil* (São Paulo: Friends of the Earth, Brazilian Amazon, 2009), p. 1.

26. Abha Chhabra et al., *Spatial Pattern of Methane Emissions from Indian Livestock*, Space Applications Centre Report (Ahmedabad: Indian Space Research Organization, 2009), p. 4; Krishi Bahawan, "Basic Animal Husbandry Statistics," Department of Animal Husbandry Dairy & Fishing, Ministry of Agriculture, Government of India, New Delhi, 2010, pp. 49–50.

27. Priya Jagannathan, "Iffco, New Zealand's Fonterra to Set Up Dairy Here," *Economic Times*, 4 October 2010.

28. FAO, *The State of the World's Land and Water Resources for Food and Agriculture (SOLAW)* (Rome: 2011); FAO, "Scarcity and Degradation of

Land and Water: Growing Threat to Food Security," press release (Rome: 28 November 2011).

29. International Panel for Sustainable Resource Management, *Assessing the Environmental Impacts of Consumption and Production: Priority Products and Materials* (Nairobi: U.N. Environment Programme, 2010).

30. Government of India, Ministry of Law and Justice, *The Constitution of India* (As Modified Up to the 1st December, 2007); The Official Law Reports of the Republic of Kenya, *The Constitution of Kenya*, Revised Edition 2010, National Council for Law Reporting with the Authority of the Attorney General; Republic of Ecuador, *Constitution of 2008*, Political Database of the Americas, Edmund A. Walsh School of Foreign Service, Georgetown University.

31. Meatless Monday, at www.meatlessmonday .com; "Lançamento da Campanha Segunda sem Carne em Curitiba," Segunda sem Carne, 13 March 2010, at www.svb.org.br.

Chapter 15. Biodiversity: Combating the Sixth Mass Extinction

1. United Nations Convention on Biological Diversity (CBD), at www.cbd.int/history; International Union for Conservation of Nature (IUCN), *The IUCN Red List of Threatened Species 2011.2*, Summary Statistics, at www.iucnredlist.org/about/ summary-statistics.

2. Figure 15–1 and data in text from IUCN, op. cit. note 1; M. Hoffmann et al., "The Impact of Conservation on the Status of the World's Vertebrates," *Science*, 10 December 2010, pp. 1503–09; Jean-Cristophe Vié, Craig Hilton-Taylor, and Simon N. Stuart, eds., *Wildlife in a Changing World: An Analysis of the 2008 IUCN Red List of Threatened Species* (Gland, Switzerland: IUCN, 2009); IUCN Species Survival Commission, "Sturgeon More Critically Endangered than Any Other Group of Species," press release (Gland, Switzerland: 18 March 2010).

3. Figure 15–2 and data in text from WWF, ZSL, and GFN, *Living Planet Report 2010* (Gland,

Switzerland: IUCN, 2010); Millennium Ecosystem Assessment (MA), *Ecosystems and Human Well-being: Biodiversity Synthesis* (Washington, DC: World Resources Institute, 2005); R. Leakey and R. Lewin, *The Sixth Extinction: Patterns of Life and the Future of Humankind* (New York: Bantam Dell Publishing Group, 1995); A. D. Barnosky et al., "Has the Earth's Sixth Mass Extinction Already Arrived?" *Nature*, 3 March 2011, pp. 51–57.

4. Secretariat of the Convention on Biological Diversity, *Global Biodiversity Outlook 3* (Montreal: 2010); MA, op. cit. note 3.

5. Gallup Organisation, Hungary, *Attitudes of Europeans towards the Issue of Biodiversity: Analytical Report, Wave 2*, Flash Eurobarometer 290, conducted for the European Commission (2010).

6. Kevin J. Gaston and John I. Spicer, *Biodiversity: An Introduction*, 2nd ed. (Hoboken, NJ: Wiley-Blackwell, 2004); CBD, op. cit. note 1.

7. J. Diamond, "Easter Island's End," *Discover Magazine*, August 1995; Australian Government, "European Wild Rabbit (*Oryctolagus cuniculus*)," *Invasive Species Fact Sheet*, Department of Sustainability, Environment, Water, Population and Communities, 2011.

8. U.N. Environment Programme (UNEP), *Towards a Green Economy: Pathways to Sustainable Development and Poverty Eradication* (Nairobi: 2011).

9. UNEP, Report of the Sixth Meeting of the Conference of the Parties to the Convention on Biological Diversity, Annex I, Decision VI/26, 2002; UNEP, Report of the Tenth Meeting of the Conference of the Parties to the Convention on Biological Diversity, 2010.

10. Box 15–1 from the following: UNEP, *Keeping Track of Our Changing Environment. From Rio to Rio+20 (1992–2012)* (Nairobi: 2011), p. 4; J. Lyytimäki et al., "Nature as a Nuisance? Ecosystem Services and Disservices to Urban Lifestyle," *Journal of Integrative Environmental Sciences*, September 2008, pp. 161–72; A. H. Petersen et al., "Natural Assets in Danish National Parks" (in Danish), Copenhagen University, 2005; B. Normander et al., "State of the Environment 2009—Part A: Denmark's Environment under Global Challenges," National Environmental Research Institute, Aarhus University, 2009; Brian McCallum and Alison Benjamin, *Bees in the City: The Urban Beekeepers' Handbook* (York, U.K.: Guardian Books, 2011); HoneyLove Urban Beekeepers, at honeylove.org; Eagle Street Rooftop Farm, at rooftopfarms.org; Greenroof & Greenwall Projects Database, at www.greenroofs.com; Windowfarms, "A Vertical, Hydroponic Garden for Growing Food in Your Window," at www.windowfarms.org; Fred Pearce and Orjan Furubjelke, "Cultivating the Urban Scene," in Paul Harrison and Fred Pearce, eds., *AAAS Atlas of Population and Environment* (Washington, DC, and Berkeley, CA: American Association for the Advancement of Science and the University of California Press, 2000).

11. J. Rockström et al., "A Safe Operating Space for Humanity," *Nature*, 24 September 2009, pp. 472–75.

12. Intergovernmental Platform on Biodiversity and Ecosystem Services, at www.ipbes.net/about-ipbes.html.

13. B. Normander et al., "Indicator Framework for Measuring Quantity and Quality of Biodiversity— Exemplified in the Nordic Countries," *Ecological Indicators*, February 2012, pp. 104–16.

14. U.N. Food and Agriculture Organization, *Global Forest Resources Assessment 2010* (Rome: 2010).

15. Guy Marcovaldi, Neca Marcovaldi, and Joca Thomé, "Retail Sales Help Communities and Sea Turtles in Brazil," in *The State of the World's Sea Turtles: SWOT Report Volume IV* (Arlington, VA: 2009), p. 35; "Forests—Investing in Natural Capital," in UNEP, op. cit. note 8.

16. IUCN/UNEP, *The World Database on Protected Areas* (WDPA), at www.protectedplanet.net; UNEP, op. cit. note 10.

17. Forest Watch Indonesia and Global Forest Watch, *The State of the Forest: Indonesia* (Bogor, Indonesia, and Washington, DC: 2002).

18. Box 15–2 from Clive Wilkinson, ed., *Status of Coral Reefs of the World: 2008* (Townsville, Australia: Global Coral Reef Monitoring Network, 2008); Alice McKeown, "One-Fifth of Coral Reefs Lost, Rest Threatened by Climate Change and Human Activities," *Vital Signs Online*, May 2009; O. Hoegh-Guldberg et al., "Coral Reefs Under Rapid Climate Change and Ocean Acidification," *Science*, 14 December 2007, pp. 1737–42.

19. D. Pauly et al., "Towards Sustainability in World Fisheries," *Nature*, 8 August 2002, pp. 685–95; "Fisheries—Investing in Natural Capital," in UNEP, op. cit. note 8.

Chapter 16. Ecosystem Services for Sustainable Prosperity

1. Lack of biophysical limits from R. Beddoe et al., "Overcoming Systemic Roadblocks to Sustainability: The Evolutionary Redesign of Worldviews, Institutions, and Technologies," *Proceedings of the National Academy of Sciences*, 24 February 2009, pp. 2,483–89; planetary boundaries from J. Rockström et al., "A Safe Operating Space for Humanity," *Nature*, 23 September 2009, pp. 472–75, and from W. Steffen, J. Rockström, and R. Costanza, "How Defining Planetary Boundaries Can Transform Our Approach to Growth," *Solutions*, May 2011, pp. 59–65.

2. Capital accumulation by the few over asset building by the many from Bureau of National Economic Accounts, "Current Dollar and 'Real' GDP," U.S. Department of Commerce, Washington, DC, 2007, and from J. E. Stiglitz, *Globalization and Its Discontents* (New York: W. W. Norton & Company, 2002); rising income disparity and ecosystem degradation from J. G. Hollender et al., "Creating a Game Plan for the Transition to a Sustainable U.S. Economy," *Solutions*, June 2010, pp. 36–41.

3. Constrain real progress from H. E. Daly, "From a Failed-Growth Economy to a Steady-State Economy," *Solutions*, February 2010, pp. 37–43.

4. Material consumption beyond real need reducing overall well-being from R. A. Easterlin, "Explaining Happiness," *Proceedings of the National*

Academy of Sciences, 16 September 2003, pp. 11,176–83.

5. Definition of ecosystem services from R. Costanza et al., "The Value of the World's Ecosystem Services and Natural Capital," *Nature*, 15 May 1997, pp. 253–60, and from Millennium Ecosystem Assessment (MA), *Ecosystems and Human Well-being: Synthesis* (Washington, DC: Island Press, 2005). Ecosystem processes and functions describe biophysical relationships and exist regardless of whether or not humans benefit. Ecosystem services, on the other hand, only exist if they contribute to human well-being and cannot be defined independently. See E. F. Granek et al., "Ecosystem Services as a Common Language for Coastal Ecosystem-based Management," *Conservation Biology*, vol. 24, no. 1 (2010), pp. 207–16.

6. Definition of natural capital from R. Costanza and H. E. Daly, "Natural Capital and Sustainable Development," *Conservation Biology*, March 1992, pp. 37–46; combining the different forms of capital from R. Costanza et al., "Valuing Ecological Systems and Services," *F1000 Biology Reports*, July 2011, p. 14.

7. Source of ecological conflict from Costanza et al., op. cit. note 6.

8. Economist valuation methods from A. M. Freeman, *The Measurement of Environmental and Resource Values: Theories and Methods*, 2nd ed. (Washington, DC: RFF Press, 2003); ecologist valuation methods from R. Costanza, "Value Theory and Energy," in C. Cleveland (ed.), *Encyclopedia of Energy, Vol. 6* (Amsterdam: Elsevier, 2004), pp. 337–46.

9. Costanza et al., op. cit. note 5.

10. MA, op. cit. note 5, p. 14; MA, *Living Beyond Our Means: Natural Assets and Human Well-Being: Statement from the Board* (Washington, DC: World Resources Institute, 2005), p. 2.

11. P. Sukhdev and P. Kumar, *The Economics of Ecosystems and Biodiversity (TEEB)* (Brussels: European Communities, 2008).

12. Imperfect information from B. Norton, R.

Costanza, and R. Bishop, "The Evolution of Preferences: Why 'Sovereign' Preferences May Not Lead to Sustainable Policies and What to Do About It," *Ecological Economics*, February 1998, pp. 193–211; quantify the amount of a given service derived from that system from E. B. Barbier et al., "Coastal Ecosystem-based Management with Non-linear Ecological Functions and Values," *Science*, 18 January 2008, pp. 321–23, and from E. W. Koch et al., "Non-linearity in Ecosystem Services: Temporal and Spatial Variability in Coastal Protection," *Frontiers in Ecology and the Environment*, February 2009, pp. 29–37.

13. Broader notion of value from R. Costanza, "Social Goals and the Valuation of Ecosystem Services," *Ecosystems*, January-February 2000, pp. 4–10.

14. Ecosystem services essential to the existence of human society from G. C. Daily et al., *Ecosystem Services: Benefits Supplied to Human Societies by Natural Ecosystems* (Washington, DC: Ecological Society of America, 1997); connection to human health from MA, op. cit. note 5.

15. Assets that should be held in common from Beddoe et al., op. cit. note 1, and from I. Kubiszewski, J. Farley, and R. Costanza, "The Production and Allocation of Information as a Good That Is Enhanced with Increased Use," *Ecological Economics*, vol. 69 (2010), pp. 1,344–54.

16. Common asset trust from P. Barnes, *Capitalism 3.0* (San Francisco: Berrett-Koehler, 2006), and from P. Barnes et al., "Creating an Earth Atmospheric Trust" (letter), *Science*, 8 February 2008, p. 724; freely available information from Kubiszewski, Farley, and Costanza, op. cit. note 15.

17. Payment for ecosystem services from J. Farley and R. Costanza. "Payments for Ecosystem Services: From Local to Global," *Ecological Economics*, vol. 69 (2010), pp. 2,060–68.

18. Ecosystem services in public media from J. D. Schwartz, "Should We Put a Dollar Value on Nature?" *Time*, 6 March 2010; C. Asquith, "Dow Chemical and The Nature Conservancy Team Up to Ask, What Is Nature Worth? Interview with Mark Weick and Michelle Lapinski," *Solutions*, vol. 2, no. 6 (2011).

19. Ecosystem Services Partnership, at www.fsd .nl/esp; Wealth Accounting and Valuation of Ecosystem Services, at go.worldbank.org/ PL08P9FTN0; Intergovernmental Platform on Biodiversity and Ecosystem Services at ipbes.net.

Chapter 17. Getting Local Government Right

1. Description of Cameroon case based on Nchunu Justice Sama, "Promoting the Foundations of Environmental Governance and Democracy," presentation at World Resources Institute, Washington, DC, 15 September 2009.

2. *Foundation for Environment and Development (FEDEV) & 1 Other v. Bamenda City Council & 2 Others*, HCB/19/08 (Mezam High Court).

3. U.N. Conference on Environment and Development, "Principle 10," *Rio Declaration on Environment and Development* (Nairobi: U.N. Environment Programme, 1992).

4. Box 17–1 from Alexandre Kiss and Dinah Shelton, *Guide to International Environmental Law* (Amsterdam: Martinus Nijhoff Publishers, 2007)

5. "Chapter 28: Local Authorities' Initiatives in Support of Agenda 21," in United Nations, *Earth Summit Agenda 21: The United Nations Programme of Action from Rio* (New York: 1993).

6. Population Division, *World Urbanization Prospects: The 2009 Revision* (New York: United Nations, 2010); Asian Development Bank, *City Development Strategies to Reduce Poverty* (Manila: 2004); Panupong Panudulkitti, *How Does the Level of Urbanization Matter for Poverty Reduction?* (Atlanta, GA: Georgia State University, 2007).

7. Thomas Tyler and David L. Markell, "Using Empirical Research to Design Government Citizen Participation Processes: A Case Study of Citizens' Roles in Environmental Compliance and Enforcement," *University of Kansas Law Review*, vol. 57, no. 1 (2007), pp. 1–38; Thomas Webler, Seth Tuler, and Rob Krueger, "What is a Good Public Participation Process? Five Perspectives from the Public," *Environmental Management*, vol. 27, no. 3 (2002), pp. 435–50.

8. Local Governments for Sustainability, *Second Local Agenda 21 Survey*, Submitted to the Commission on Sustainable Development (New York: United Nations, 2002); Isabel M. Garcia-Sanchez and Jose-Manuel Prado-Lorenzo, "Determinant Factors in the Degree of Implementation of Local Agenda 21 in the European Union," *Sustainable Development*, vol. 16, no. 1 (2008), pp. 117–34; Paul Selman, "Local Agenda 21: Substance or Spin?" *Journal of Environmental Planning and Management*, vol. 41, no. 5 (1998), pp. 533–53.

9. Box 17–2 based on The Access Initiative website, at www.accessinitiative.org/about.

10. Table 17–1 from The Access Initiative collaborative partners, which includes the Bolivian Society for the Defense of Nature (PRODENA), The Center for Human Rights and Environment (CEDHA), CoopeSolidar (Costa Rica), The Ecuadoran Center for Environmental Law (CEDA), Environmental Management and Law Center (Hungary), the Foundation for Development and Environment (Cameroon), the Mexican Center for Environmental Law (CEMDA), Participa (Chile), and the Thailand Environment Institute. OMB Watch (United States), which is not a member of The Access Initiative, also contributed to the research.

11. Equity and Government Accountability Project, at www.ombwatch.org/EGAP.

12. Research provided by Sofia Plagakis, OMB Watch.

13. Case studies submitted to the Access Initiative.

Index